SNAKES
AND OTHER REPTILES
OF SOUTHERN AFRICA

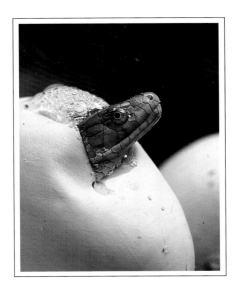

BILL BRANCH

Dedication
To Dove – a first small gift for her love and understanding

Published by Struik Nature
(an imprint of Random House Struik (Pty) Ltd)
Reg. No. 1966/003153/07
80 McKenzie Street, Cape Town, 8001 South Africa
PO Box 1144, Cape Town, 8000 South Africa

Visit us at **www.randomstruik.co.za**
Log on to our photographic website
www.imagesofafrica.co.za
for an African experience

First Published 1993
Second Edition 2001

10 9 8

Editor: Lindsay Norman
Designer: René Greeff
Page layout: Joan Sutton
Typesetter: Suzanne Fortescue, Struik DTP

Reproduction by Hirt & Carter Cape (Pty) Ltd
Printed and bound by Times Offset (M) Sdn Bhd

ISBN 978 1 86872 619 6

Front cover: *Telescopus semiannulatus*
Back cover: *Bitis caudalis*

CONTENTS

Key to symbols used in this book

🕱   Snakes whose bite is fatal to humans

☺   Snakes whose bite is poisonous but non-fatal to humans

*E   Endangered (Those species in danger of extinction and whose survival is unlikely without the implementation of special protective measures.)

*V   Vulnerable (Those species likely to move into the endangered category if environmental disturbances continue.)

*Re   Restricted (Those species with a restricted distribution in South Africa but whose main distribution falls outside the political boundaries of the area.)

*P   Peripheral (Those species with a restricted distribution in South Africa but whose main distribution falls outside the political boundaries of the area.)

*R   Rare (Those species not at present endangered or vulnerable but occurring in such small numbers and/or in such a restricted or specialized habitat that they are at risk.)

Introduction

With increasing public concern for the environment, interest in less fashionable wildlife such as reptiles and amphibians has grown. If we are to return to harmony with our world, and to integrate with the biosphere rather than continually exploit it, then we need a deeper understanding of life around us. This book is designed to introduce our reptiles, from the feared snakes and crocodile, to the ever-popular tortoises and the secretive and neglected lizards.

Southern Africa, that region south of the Zambezi and Cunene rivers, has a tremendous variety of habitats. These harbour a very diverse reptile fauna of nearly 500 species. More than half of these are endemic to the region, that is, found nowhere else. Unfortunately, not all of these can be included in this book. Those chosen emphasize the more colourful and conspicuous species, and those unique to or endangered in southern Africa. However, a number of rare and localized species have also been included to reflect the great diversity which the region supports.

Most people realize that not all snakes are poisonous, but many still kill them anyway. The lazy argument is that "it is better to be safe than sorry". However, of 146 snake species found in southern Africa, less than one quarter are dangerous. Of these, 14 have caused death and a further 18 have venoms whose symptoms may range from inconvenience to serious. Many regions have less than a handful of dangerous species, and it requires little effort to recognize these. All are included in this book.

Living reptiles are an extremely diverse class, and the relationships between the main orders and families are still controversial. Four main lineages survive. The lizard-like tuataras (Order Rhynchocephalia) are restricted to New Zealand. The remaining three orders all occur in southern Africa, including the crocodilians (Order Crocodylia), tortoises, terrapins and turtles (collectively called chelonians, Order Chelonia), and snakes, lizards and worm lizards (collectively called squamates, Order Squamata). Crocodiles are more closely related to dinosaurs and birds than to other reptiles. Squamates (scaled reptiles) are all closely related, with snakes and amphisbaenians (worm lizards) evolving from lizards approximately 80 million years ago.

Unlike birds, which are often common and highly visible, most reptiles are small, secretive and shy. To find or observe them usually requires patience and careful searching. Reptiles have only limited mobility and many have very specific habitat requirements. In general, lizards can be considered habitat-linked and snakes food-linked. This is reflected in many of their common names. Thus, among snakes we talk of egg eaters, centipede eaters and slug eaters, and among lizards of desert lizards, rock lizards and the water monitor. In practical terms this means that lizards often have very small ranges within which they inhabit specific places. Snakes often range over large areas and occur in different habitats, but search for specific prey.

In the southern, more temperate regions, mating occurs in spring as daytime temperatures rise, whilst in the north mating often occurs at the start of the wet season. Male snakes become more active at these times, searching for receptive females. Dominant males among the colonial agamas and skinks develop bright breeding colours, and lead active lives defending their territories and chasing mates.

The majority of reptiles, including all crocodiles and chelonians, are oviparous (egg-laying). Most lay clutches of between five and 20 eggs, although large sea turtles may lay up to 1 000 in a season; most geckos lay only two eggs at a time. With two exceptions, parental care in all local reptiles ends when the eggs are laid and the nest hole covered. The Southern African Python and Nile Crocodile both brood their eggs until they hatch. In crocodiles and sea turtles the sex of the embryo is dependent upon the temperature at which the egg is incubated. In crocodiles, males develop in eggs at high temperatures, while the same temperatures in turtles produce females. Many squamates are viviparous, retaining their eggs within the body and giving birth to live young. This is usually associated with life in cool climates.

How to use this book

Clarity and ease of use have been the main criteria in the design of this guide. In the succinct species descriptions, key features for identification are emphasized in *italic* type. In some difficult groups, such as thread snakes and sand lizards, it may be necessary to have the specimen in hand to obtain detailed scale counts before a positive identification can be assured.

As with birds, field identification is not always easy and depends on sensible and good observation. When first spotting a reptile, try to note the following features. What was the general build? Was the coloration uniform or patterned, and, if the latter, was it plain, striped, blotched, etc? Was the tail longer or shorter than the body? Were the body scales large or small, smooth or rough, arranged in rows or scattered? What habitat was it in, and what was it doing? These features should be jotted in a field book, as an aid for comparison with the pictures in the species accounts. The thumbnail silhouettes allow quick access to the reptile group most closely resembling the species you have seen. Further field hints are given below. A glossary of specialized terms and diagrams showing the anatomy of various reptiles are featured on pages 139-140.

When you find an illustration that is similar to your unknown species, check whether it occurs in the region. If there is a big difference between where you have found it and where it should be, try another picture. Although many reptile distributions are poorly known, you are unlikely to discover major range extensions. It is more likely that you have simply misidentified it. Remember that many reptile species, particularly lizards, are not included in this guide. You are directed to a selection of further reading (page 140) that should facilitate identification of any unusual species.

Field hints

In most of southern Africa, provincial legislation prohibits the collection, transport or possession of reptiles and amphibians. It is therefore illegal to catch them unless special permission has been obtained. It is also necessary to obtain permission to walk or collect on private land. Given these restrictions, however, nothing prevents you from observing and enjoying reptiles in the wild. Bird-watching is a pastime enjoyed by thousands of people in southern Africa, and there is no reason why watching reptiles cannot be as enjoyable.

The first requirement for field work is that you wear suitable clothing, that is, subdued in colour and robust enough to survive thorns and sharp rocks. A good pair of compact binoculars that focus down to 2-3 meters is also useful, as well as a small pocket-sized notebook to record observations. Walk slowly and quietly, scanning the ground or rock outcrops ahead. Reptiles are ectotherms and thus gain their heat externally, usually from basking in the sun. On cool days or in winter they are rarely active. Being small they quickly warm up, and then usually retreat to shade where they avoid over-heating and are less visible to predators. On hot days, they are therefore best observed in the early morning and late afternoon. On overcast days they may be active throughout the day.

Most lizards and tortoises have small home ranges in which they have one or several retreats that they use all their lives. The "fright" distance, that is, the closest approach they will allow an observer before retreating into their shelter, is usually relatively small and well within the close range of normal binoculars. Moreover, after several visits they often become habituated to the presence of an observer, and behave normally. You can then watch interesting territorial and mating displays.

Identification in the field

Snakes
It is not difficult to distinguish the main groups of snakes in the field. Differences in coloration, behaviour and habitat all help to make a general identification (ID) possible. Precise identification may require the capture of the specimen for detailed scale counts. The main groups can be distinguished by the following general features.

Burrowing primitive snakes (pages 8-11)
Head blunt, with vestigial eyes; body shiny, and tail short; belly scales not enlarged. Usually plain black or brown in colour. Burrow underground; wriggle but never bite.
Pythons (pages 12-13)
Head triangular, covered with small scales, and with prominent pits on the lips; body strong and muscular with smooth, small scales; belly scales enlarged; tail moderately long. Usually blotched in brown, cream or olive. Active in the early morning or evening.
Typical snakes (pages 14-15)
Head covered with large, symmetrical scales on crown; belly scales enlarged. Body form, coloration, and habits very varied.
Cobras and mambas (pages 51-56)
Head covered with large symmetrical scales on crown; belly scales enlarged; large front fangs. Body form, coloration, and habits very varied. Except for garter snakes, all threaten by raising the forebody and inflating or spreading the neck.
Adders (pages 57-62)
Head triangular, covered with small scales (except night adders), and usually with a prominent mark (V-shaped or arrow-shaped) on the crown; body fat, with rough, small scales on top; belly scales enlarged; tail very short. Usually blotched in browns and greys. Active in the early morning or evening. Usually give a warning hiss, and strike readily.

Snakes lack limbs, and have evolved from lizards. However, there are also lizards which lack limbs, and it is not always easy to distinguish these legless lizards from snakes. Most snakes have enlarged belly scales (ventrals) that aid locomotion; no legless lizard has these. Snakes' eyes, when present, lack eyelids and have an unblinking state; many legless lizards retain eyelids. However, burrowing snakes and lizards require neither eyes nor enlarged ventral scales, and many have lost both. Both groups therefore look very similar, and there is no simple rule for telling them apart.

However, all blind snakes have very short tails with a sharp spine at the tip. They also have rounder, blunter heads than any of the legless lizards.

Lizards
Lizards cause the most confusion to the casual observer. Fortunately they have family characteristics (a "jizz" to bird-watchers) that allow a general ID.

Skinks (pages 63-74)
They have smooth, shiny, cylindrical bodies, with narrow pointed heads, no obvious neck, and tapering tails that are not much longer than the body. They move slowly, searching for food, and when disturbed usually slink behind cover, keeping a watchful eye on danger. Many burrowing species have short tails, and usually have vestigial or no limbs.

Lacertids (desert and sand lizards, etc; pages 75-84)
Easily confused with skinks, as they have a similar body shape. However, they have granular, non-shiny body scales, and the tail is usually noticeably longer than the body. They are mostly terrestrial, 'sit-and-wait' predators, that catch food by quickly dashing from cover. When disturbed they sprint rapidly from bush to bush.

Plated lizards (pages 85-88)
Large, terrestrial lizards, they have cylindrical, shiny bodies with long tails. A prominent lateral fold on the side of the body distinguishes them from skinks. They move slowly, searching for food, which is often uncovered by scratching with the forelimbs. They often 'toboggan' down slopes on their shiny bellies.

Girdled lizards (pages 89-98)
Live mainly in rock cracks. They have flattened bodies, covered in rings of spiny scales. The head is triangular with a narrow neck, and the tail is usually spiny. They bask, head up, on prominent rocks, and retreat into a rock crack when disturbed. Unlike agamas, they have large, symmetrical scales on top of the head.

Monitors (page 99)
Very large lizards; even as hatchlings they are bigger than most other lizards. The snake-like tongue is constantly flicked in and out, and is a good field feature.

Agamas (pages 100-103)
Medium-sized lizards with fat, spiny bodies, a narrow neck, and well-developed legs. They often perch in prominent spots with the head held high.

Chameleons (pages 104-107)
Slow-moving, mainly arboreal species. Their unusual clasping feet, telescopic tongue, and protruding eyes that move independently, are all unique.

Geckos (pages 108-125)
Unlike other lizards, they are mostly nocturnal. Day geckos all climb on rocks or trees, and can leap across small gaps. They usually have flared toe tips. When disturbed they hide in rock cracks or behind branches. The small, granular scales are not shiny.

Chelonians
Tortoises, terrapins and turtles are characterized by their protective shell. They have different habits, and are distinguished by their feet. Tortoises live on land, and have thick short feet. Terrapins live in freshwater, and have a webbed frill to the hindlimbs. Turtles are marine, returning to land only to lay their eggs. The forelimbs are flippers and cannot be withdrawn into the shell.

BLIND SNAKES (Family Typhlopidae)

These primitive snakes have *no teeth in the lower jaw*. The cylindrical body is covered in small, smooth, overlapping scales and they *do not have enlarged belly scales*. The *vestigial eye* consists of a small black spot beneath the head scales. The *very short tail ends in a spine*. They spend all their life burrowing underground, and feed on ant and termite larvae. They are found throughout much of the tropical region, with 8 local species.

Bibron's Blind Snake *(Typhlops bibronii)* 35-40 cm

A stout blind snake, with *30 scales around the body* and an *angular snout*. The body is plain brown, sometimes olive-brown, with a lighter belly. It is restricted to the highveld and coastal grasslands of South Africa, with a relict population in Zimbabwe. From 5-12 thin-walled eggs are laid in summer.

Delalande's Blind Snake *(Rhinotyphlops lalandei)* 25-30 cm

This pink-grey blind snake can be distinguished from other blind snakes by its more *slender build* and prominent *horizontal edge to the snout* which is used for burrowing. There are *26-30 scales at midbody*. It is restricted to the eastern regions of the subcontinent, with scattered records in Namibia. Females lay 2-4 eggs.

Schlegel's Blind Snake *(Rhinotyphlops schlegelii)* 60-80 cm

This thick-bodied snake is the largest typhlopid in the world. It is distinguished by the *30-44 scales at midbody* and a prominent *horizontal edge to the snout.* Coloration is variable, and the body may be plain, blotched or finely striped. When skin is shed the body is bright blue-grey with dark markings, but the skin tans with time to a rich red-brown

Newly shed

that matches the soil colour. Very large specimens are only seen when forced to the surface by floods. It lives deep underground, crawling into the brood chambers of termite nests and eating the larvae. Large fat stores allow it to undertake long fasts. It lays large numbers of eggs which take 5-6 weeks to hatch.

Flower-pot Snake *(Ramphotyphlops braminus)* 14-15 cm

C. MATTISON

A very small, slender blind snake that has a *rounded snout, 20 midbody scale rows*, and between *300 and 350 scales along the backbone*. It is uniform grey to pale brown, with a lighter belly and *cream blotches on the snout and anal region*. It is an Asian species, probably introduced to Cape Town early in the last century. Now also known from Durban and Beira in Mozambique. It is a self-fertilizing, all-female species that lays 2-6 minute (2 x 6 mm) eggs. It is often transported in nursery plants, hence its common name.

THREAD SNAKES (Family Leptotyphlopidae)

These primitive snakes are the smallest in the world. The very thin, cylindrical body is covered in small, smooth scales. Like blind snakes, they *lack enlarged belly scales* and have *vestigial eyes*. They have *no teeth in the upper jaw*. All local species have blunt heads, and the short tail is relatively longer than that of blind snakes. They live underground and follow the chemical trails of ants and termites to their nests where they eat the larvae and defenceless workers, swallowing small prey whole and biting off the abdomens of large workers. Found in Africa, Asia and South America, with nine local species. Most species lay a small number of elongate eggs which are joined together like a string of sausages.

Distant's Thread Snake *(Leptotyphlops distanti)* 13-20 cm

A *thin, grey-black* snake. When dry, the scales may turn silvery in colour and become pale-edged. The *very broad rostral scale is fused with the prefrontal*, and is *more than half the width of the head* at the level of the rear border of the eye. The *occipital scales are divided*. It is mainly restricted to the Lowveld. Breeding unknown.

Long-tailed Thread Snake *(Leptotyphlops longicaudus)* 18-22 cm

A slender thread snake which has only 10 midbody scale rows and a *very long tail* (34-58 subcaudals). The *prefrontal scale is separated from the rostral*. The body is uniform lilac to dark pink-grey above, with a fleshy-pink belly. Breeding unknown.

Western Thread Snake *(Leptotyphlops occidentalis)* 18-20 cm

 This very thin thread snake is the largest local species, and usually has *more than 300 scales along the backbone*. It is grey-brown to purple-brown in colour, and pale edges to the scales may create a chequered effect. It is mainly restricted to the rocky deserts of Namibia. It may crawl into cracks on rock outcrops as it searches for ant nests. This snake is only likely to be encountered at the surface on summer nights. Breeding unknown. *P

Peter's Thread Snake *(Leptotyphlops scutifrons)* 18-24 cm

 The most widespread thread snake in southern Africa, extending from Tanzania to KwaZulu-Natal. It can be distinguished by its *short tail* (19-39 subcaudals) and *wide rostral* (about a third the width of the head at the level of the eye). The tail ends abruptly in a spine. The body is uniform black, but may turn silvery when dry. It is usually found under logs or stones. From 3-7 elongate eggs are laid in summer. When handled it may turn limp and sham death.

PYTHONS (Family Boidae)

A diverse family, found throughout tropical regions and containing the viviparous boas and egg-laying pythons. It includes the world's largest snakes, the Anaconda (*Eunectes murinus*) of the Amazon basin, and the Reticulated Python (*Python reticulatus*) of Asia. Of the 4 species found in Africa, 2 occur in southern Africa.

Southern African Python (*Python natalensis*) 300-500 cm

Adult

Female with eggs

Africa's second largest snake. The stout body with very *small, smooth scales in 78-95 rows at midbody* make it unmistakable. The *triangular head* is covered in *small, irregular scales*, and the *upper lips have only 2 heat-sensitive pits* on each side. The body is blotched and there is a large, dark *spearhead mark on the crown*. It favours rocky or bushy areas, usually close to water. Small mammals form the main diet, although large adults may tackle small antelope. Up to 100 eggs are laid in a hollow tree, termite nest or antbear hole. The female broods the eggs, coiling around them and shivering to generate heat if the eggs get cold. Although not poisonous, pythons bite readily and large individuals should be treated with caution. *V

Hatchling

Adult

This rare species can be distinguished from the larger Southern African Python by having only *57-61 rows of scales at midbody*. The *triangular head* is also covered with *small tubercles* and the *upper lip has 5 heat-sensitive pits on each side*. The body is pale red-brown, with scattered black-edged white spots and bands. It lives in the rugged, arid mountains of northern Namibia and Angola. Small birds, and occasionally rodents, form the main diet, and 5-6 large eggs are laid in summer. This gentle snake rarely bites.

TYPICAL SNAKES (Family Colubridae)

A very diverse group that includes nearly 2 000 species, distributed throughout tropical and temperate regions. The *head usually has large, symmetrical scales*. This group is well represented in the region, with over 85 species found here.

Common Brown Water Snake *(Lycodonomorphus rufulus)* 60-80 cm

This gentle, inoffensive species is the most common water snake in southern Africa. The *eyes have elliptical pupils* and are set on the sides of the small head. The *first upper labial lacks a backward projection*. The *plain olive body has smooth scales*. The *upper lip is not spotted* and, like the belly, is pale yellow-pink in colour. In many individuals the *longish tail (53-86 subcaudals)* is darker below. It hunts at night, and is frequently found under cover around water margins. From 6-10 eggs are laid in late summer.

Dusky-bellied Water Snake *(Lycodonomorphus laevissimus)* 70-80 cm

The largest water snake in the region. The *small eyes have round pupils* and are set high on the sides of the small, flattened head. The *first upper labial has an obvious backward projection*. The back is uniform olive to brown-black, sometimes with a pale stripe on the lower flank. The belly is cream to yellow, with a *broad central dark*

band. The *upper lip is usually spotted*. An aquatic species, often found submerged in quiet backwaters and pools, where it searches for prey during the day among rocks and sunken logs. Tadpoles, frogs and fish form its main diet. Up to 17 eggs are laid in summer. Bad-tempered and ever ready to bite, it may also void a foul-smelling cloacal fluid when annoyed.

Floodplain Water Snake *(Lycodonomorphus obscuriventris)* 50-60 cm

A rare East African species that just enters the region along the Mozambique floodplain, reaching as far as Swaziland. It is a small snake, easily confused with the Common Brown Water Snake. However, the body is dark olive to blackish and the *upper lip has a prominent yellow stripe*. The orange-yellow belly is sometimes faintly spotted. The *short tail (37-52 subcaudals)* is dark below. It is a secretive snake and hunts for frogs and tadpoles in the late afternoon or early evening. A gentle, shy species that is most likely to be encountered foraging in the wet margins of small vleis and streams. Its breeding biology is unknown.

Swazi Rock Snake *(Lamprophis swazicus)* 50-70 cm

An unusual snake, with a long, slender body, and a *distinct and flattened head with large bulging eyes*. There are only *17 midbody scale rows*. The body is *uniform light to dark red-brown* with a creamy-white belly. It is endemic to the eastern Mpumalanga highlands and western Swaziland, where it inhabits rock outcrops. During the day it is most likely to be found sheltering under rock flakes. It hunts at night, when its large eyes with ventrical pupils help it to find sleeping lizards. A small number of elongate eggs are laid in summer. *R

15

Southern Brown House Snake *(Lamprophis capensis)* 60-100 cm

Found throughout the region, this is the largest house snake. It has an off-white belly and rust-red body that may become *almost black in old individuals*. It is easily distinguished from all other local snakes by the *pair of thin yellow stripes on the side of the head*. In some individuals these continue onto the forebody. Terrestrial and nocturnal, it feeds mainly on rodents, although in desert regions it also eats lizards. It is tolerant of urban conditions and occurs commonly in towns. Up to 18 eggs are laid in summer.

Olive House Snake *(Lamprophis inornatus)* 60-100 cm

A large, thick-bodied house snake, restricted to the moister coastal regions of the southern Cape and KwaZulu-Natal, extending along the eastern escarpment. It is *uniform olive-green* in colour, with a light grey-green belly. It is distinguished from olive-coloured water snakes by having *21-25 midbody scale rows* and no patterning on the belly or beneath the tail. Rodents form the main diet, although it also eats frogs and even other snakes. From 5-15 eggs are laid in summer. Usually peaceful, it settles well into captivity.

Aurora House Snake *(Lamprophis aurora)* 50-80 cm

Adult

This beautiful, gentle house snake is unfortunately rare. Of similar coloration and distribution to the Olive House Snake, it can be distinguished by the *prominent orange-yellow stripe along the backbone*. The young sparkle, each scale having a pale yellow bar. This imparts a *speckled appearance*, which could cause confusion with the spotted phase of the harlequin snake (see page 40). However, it lacks front fangs and is completely harmless. Nocturnal, it feeds mainly on nestling rodents, although lizards and frogs are also eaten. From 8-12 eggs are laid, and the hatchlings are about 20 cm long.

Juvenile

Spotted House Snake *(Lamprophis guttatus)* 40-60 cm

Although rarely seen, this small, slender house snake is common in rock outcrops of the Karoo and eastern escarpment mountains. The *head is flattened with large eyes*. Coloration is varied, but the *body is always blotched*. In the western Cape the body is light brown or tan with alternating or paired dark brown blotches on the forebody. In KwaZulu-Natal and Mpumalanga the body is pinkish to silvery-grey with large, pale-edged, purple-brown blotches that extend along the whole body and may fuse to form a zigzag. It is a shy species that shelters in rock cracks during the day and hunts at night for lizards. Up to 8 eggs are laid in summer.

17

Fisk's House Snake *(Lamprophis fiskii)* 25-35 cm

 Few local snakes are as beautiful, or as rare, as this diminutive house snake. The *lemon yellow body* has a double row of alternating *dark brown blotches*, that may fuse to form a zigzag pattern. The belly is creamy white. The *rounded head* has large eyes, with vertical pupils, whilst the tail is relatively short (28-34 subcaudals). Fewer than 20 specimens are known, from widely scattered localities in the Karoo and Little Namaqualand. It is probably mostly underground, emerging at night to feed. When annoyed it hisses and tightly coils and uncoils the front and rear of its body. Breeding unknown. *R

Yellow-bellied House Snake *(Lamprophis fuscus)* 40-60 cm

 Few people have seen this species, although it is widely distributed in the eastern regions. It is a *slender, round-headed* house snake, with a *relatively long tail* (56-74 subcaudals). The back is plain and dark to light olive, whilst the *belly, flanks and upper lip are characteristically bright yellow*. There are only *19 midbody scale rows*. It spends most of its life underground, particularly in old termite nests where it feeds on nestling rodents and possibly small geckos using the nests. A small clutch of elongate eggs is laid in summer. It has a gentle disposition but does not settle well into captivity. *R

Cape Wolf Snake *(Lycophidion capense)* 30-50 cm

 Wolf snakes are named after their long, recurved teeth. Nonetheless, they are harmless, gentle snakes. They are peculiar to Africa and are found all the way to Somalia. This species is small with a flattened head hardly distinct from the body. The *eyes are small with vertical pupils*, and the *first upper labial contacts the postnasal*. The body is *uniform grey-black* (browner in the southern regions), often with each scale white-tipped. It shelters by day, emerging at dusk to search for prey. The diet consists mainly of diurnal lizards such as skinks and sand lizards. From 3-8 eggs are laid in summer.

Pygmy Wolf Snake *(Lycophidion pygmaeum)* 20-25 cm

The small, grey-blue to purple-brown *body appears faintly speckled* due to the pale-edged scales. There is a characteristic *pale edge to the blunt head, from the snout to the temporal region*. The dark belly has pale-edged ventrals. Restricted to the eastern coastal regions this rare, terrestrial species shelters in grass tussocks. Lays 3-4 eggs. *P

19

Variegated Wolf Snake (*Lycophidion variegatum*) 30-40 cm

In this species the speckles are much more extensive than those of other wolf snakes, and may fuse to cover the body in a *coarse white mottling*. The body is *more elongate* (185-204 ventrals), and the first upper labial is separated from the postnasal. It has a restricted distribution, occurring in the eastern mountains from the Lebombo Range in northern Zululand to Zimbabwe. Rock-dwelling, it can be found sheltering beneath stones or under rock flakes. Sleeping skinks and geckos form the main diet, and a few elongate eggs are laid in summer. *P

Namibian Wolf Snake (*Lycophidion namibianum*) 30-40 cm

Although easily confused with the Cape Wolf Snake, this species differs in having a *reddish to dark brown back*, heavily speckled with white, and a *broad white band on the lower flanks*. There is also usually a dark brown stripe along the middle of the belly. Further, unlike the Cape Wolf Snake the first upper labial is separated from the postnasal. It has mostly been collected from rocky areas in bushy scrubland, although one was found trying to climb up a dune. Nothing is known of its diet or reproduction, although these are likely to be similar to those of other wolf snakes.

Cape File Snake *(Mehelya capensis)* 100-150 cm

 An unusual and rarely seen snake, easily recognized by its thickset *triangular body* and *very flat head*. The almost conical, *strongly keeled scales* are separated by bare, pink-purple skin, and the *scales along the backbone are white, enlarged and have 2 keels*. The body is grey-brown with an ivory-cream belly and flanks. It is a formidable predator of other snakes, which it kills by constriction. It will even tackle venomous snakes, including cobras, to whose venom it is immune. A clutch of 5-13 relatively large eggs is laid in leaf litter in summer. Docile in disposition, it never bites. Its only offensive behaviour is to void its bowels when handled.

Black File Snake *(Mehelya nyassae)* 40-60 cm

 The flattened head, triangular body and pinkish skin between the body scales confirm that this is a file snake. It can be distinguished from the Cape File Snake by its *uniform purple-black colour*, although the belly may be creamy-white. In addition, it has a slightly *longer tail* (51-77 subcaudals) and fewer (less than 190) ventrals. It inhabits mainly savanna, but may enter coastal forest. Although small snakes are eaten, lizards, particularly skinks, form the main diet. Up to 6 eggs are laid in summer. A small, shy snake which rarely bites and usually moves jerkily in the hand.

Variegated Slug Eater *(Duberria variegata)* 30-35 cm

 Very similar in habits and appearance to the Common Slug Eater, this small snake has a *more prominent snout* and fewer ventrals (91-110). The *olive-brown to dark-brown back has 3 rows of blackish spots* that may fuse to form irregular crossbands. The dirty-yellow belly has dark reticulations. It is restricted to coastal dune vegetation in northern Zululand and adjacent Mozambique, where it feeds exclusively on slugs and snails. From 7-20 young are born in late summer. Although it may wriggle in the hand, it does not form a tight spiral to hide its head like the Common Slug Eater.

Common Slug Eater *(Duberria lutrix)* 30-40 cm

 A *stout-bodied* little snake, with a *small head*. The back is brick-red to pale brown, sometimes with a broken black line along the backbone. The paler flanks vary from grey to light brown, and the belly is cream, edged with a dark dotted line. A gardener's friend, this small, shy snake feeds entirely on slugs and snails, which it finds by following their slime trails. It is usually found hiding in damp situations, such as grass roots, rotting logs, or compost heaps, but may move around on humid, overcast days. From 6-9 young are born in late summer. It may roll into a tight spiral when handled (hence the Afrikaans name 'tabakrolletjie') and release an unpleasant cloacal fluid.

Mole Snake (*Pseudaspis cana*) 100-150 cm

Adult

Juvenile

 A widespread, large snake, easily recognized by its thick body, *slightly hooked nose*, and *small eyes with round pupils*. The body scales are usually smooth, but sometimes keeled in black snakes from the south-western Cape. Colour varies with region and age. Juveniles are light brown with 4 rows of dark, pale-edged spots. These usually fade in subadults (about 1 m), but may persist. Adults are uniform light to red-brown, sometimes grey-olive to dark brown, and usually jet black in the south-western Cape. Males have thicker, longer tails than the females. The Mole Snake is a useful constrictor that lives underground, hunting rodents and moles. A large litter of 20-40, exceptionally 95, young are born in late summer. Although not poisonous it can give a deep and painful bite. During the breeding season males become aggressive. They may bite each other and leave deep wounds.

23

Sundevall's Shovel-snout *(Prosymna sundevallii)* 24-30 cm

 Shovel-snout snakes are easily recognized by the *angular edge to the snout*. In this small species the *'shovel' is also upturned*. The body is cylindrical with smooth scales. The *short tail ends in a spine*. The pale to dark brown body has a paired row (in the southern and central regions) or a single row (northern and eastern regions) of dark spots. The belly is white. The 'shovel' helps it to tunnel in soft soils as it searches for reptile eggs, which form the sole diet. A small clutch of 3-5 elongate eggs is laid in summer.

Two-striped Shovel-snout *(Prosymna bivittata)* 26-30 cm

 Although very similar in appearance to Sundevall's Shovel-snout, this species can be distinguished by the *prominent broken orange stripe that runs along the backbone*. The rest of the body is purple-brown and the belly is white. It is usually found under stones or rotting trees on sandy soil in the arid savanna of the Kalahari region. There is an isolated population along the lower Orange River. It feeds on reptile eggs. Fat stores allow it to survive long fasts between breeding seasons. Up to 4 elongate eggs are laid in summer.

South-western Shovel-snout (*Prosymna frontalis*) 30-40 cm

 The largest and most slender shovel-snout in the region. The *angular snout is not upturned* and the fused internasals form a *single band behind the snout*. The *relatively long tail has 32-54 subcaudals*. The body is light brown to chestnut above, and a dark edge to each scale creates a striped or stippled effect. A *broad, dark-brown collar* covers the neck and fainter crossbars may occur on the forebody. The belly is white. It inhabits the western arid regions, where it is usually found under stones on sandy soil. Breeding unknown. *P

East African Shovel-snout (*Prosymna stuhlmannii*) 24-28 cm

 A small shovel-snout with *smooth scales*. The *snout is angular, but not upturned*. The *short tail* has only 17-39 subcaudals. The scales of the *dark-brown to metallic blue-black body* may be pale-centred, and paired small white spots may flank the backbone. Although usually white, the belly may be brown-black. It is a species of wooded savannas found in the north-eastern regions, and reaching as far south as northern Zululand. From 3-4 elongate eggs are laid in leaf litter in summer. Shy and secretive, it never bites or wriggles violently.

25

Cream-spotted Mountain Snake *(Montaspis gilvomaculata)* 30-45 cm

This rare snake was only described in 1991 and is known from just 4 specimens. All have been found alongside streams and marshes on the upper slopes of the KwaZulu-Natal Drakensberg. It is a small snake with a robust, short, cylindrical body, a head only slightly distinct from the neck, and a tail of moderate length. The body is shiny black-brown, with *prominent cream spots on the lip scales and throat*. Although it has *two enlarged back fangs* it is not known to be dangerous. It feeds on small frogs. Breeding unknown.

Olive Marsh Snake *(Natriciteres olivacea)* 35-50 cm

This small, plain olive to brown-coloured snake has *smooth scales in 19 rows at midbody* and a longish tail. It is usually found sheltering under logs or stones in the margins of vleis and pans in the northern savannas. Small frogs and fish form the main diet, although it has also been observed to eat winged termites. When first caught it may spin widely, and if the tail is held it breaks easily and cannot be regenerated. It is a gentle species that tames well in captivity. Up to 8 eggs are laid in early summer.

Forest Marsh Snake (*Natriciteres variegata*) 30-40 cm

 A small, thickset snake with *smooth scales*. The tail is relatively long (*60-84 subcaudals*) and may be shed if grabbed, although it cannot be regrown. The *dark olive to chestnut-brown back* usually has a broad, *darker band down the backbone*. This is usually bordered by a row of white dots, and a faint yellow collar may be present. It is found in wet montane and lowland forest, and occurs in isolated populations in northern Zululand and eastern Zimbabwe. This species frequents dead logs and rotting vegetation at forest fringes, where it usually hunts frogs. Up to 6 eggs are laid in summer. *P

Striped Swamp Snake (*Limnophis bicolour*) 45-55 cm

 A small, inconspicuous snake with a cylindrical body and small head. It can be distinguished from the marsh snakes by having a *single, triangular internasal scale* on top of the snout. The largish eyes have round pupils. The *body has a striped appearance* due to 3-4 black-edged scale rows on the flanks. The bright *belly is yellow to brick red*. Shy and secretive, it hunts small frogs and possibly small fish in the marshes of the Okavango and Zambezi River valleys. Little is known of its reproduction; a large female contained 5 eggs.

27

Western Keeled Snake *(Pythonodipsas carinata)* 45-60 cm

 Superficially viper-like, this rare, nocturnal snake is easily recognized by its *long, thin body and fragmented head scales*. The *flat head* has a distinct neck and large eyes with *vertical pupils*. The *swollen nostrils* are placed on top of the snout. The back has various pastel colours, and a double series of dark-edged blotches that may fuse into a zigzag or irregular crossbands. Although it has *large back fangs* and bites readily, it is harmless. Found in rocky desert where it feeds on small lizards and rodents. It is believed to be oviparous.

Many-spotted Snake *(Amplorhinus multimaculatus)* 45-60 cm

 An unusual, secretive snake that is found in isolated populations in the cool, moist eastern regions. It may be locally common. The *small head* has medium-sized eyes with *round pupils*. There are *17 midbody scale rows* and a *longish tail*. Coloration is variable. The back is usually green to olive-brown with a series of dark blotches and sometimes a pale dorsolateral stripe. Scattered, pale-edged scales may give a flecked appearance. It has *grooved back fangs*, and catches frogs in the early evening in waterside vegetation. From 4-12 young are born in late summer. The venom is considered harmless.

Rhombic Skaapsteker *(Psammophylax rhombeatus)* 80-120 cm

Despite its reputation this snake is harmless, its bite being less dangerous than a bee sting. The smallish head has a *rounded snout*, there are *17 midbody scale rows*, and the *tail is moderately long* with 60-84 sub-caudals. The *yellowish-brown back has 3-4 rows of dark-edged blotches* that may fuse to form an irregular zigzag or stripes (particularly in KwaZulu-Natal). It hunts during the day in moist grasslands, and has a varied diet which includes mice, frogs and lizards. Up to 30 eggs are laid in a hole in summer.

Striped Skaapsteker *(Psammophylax tritaeniatus)* 60-80 cm

Slightly smaller, but similar in build to the Rhombic Skaapsteker, this attractive snake can be distinguished by its more *pointed snout* and *shorter tail* (49-69 sub-caudals). It has a *prominent striped pattern*, with 3 black-edged dark brown stripes on a pale grey body. The middle stripe may be divided by a fine yellow line. The upper lip and belly are plain white. It hunts small mice in the northern savannas, but will also take frogs and lizards. Up to 18 eggs are laid in a hole in summer, but are not guarded by the female. Gentle and inoffensive, it may wriggle in the hand, but never bites.

Olive Grass Snake *(Psammophis mossambicus)* 100-140 cm

A big, robust snake, with a *long tail* and *large scales in 17 midbody rows*. There are usually *more than 164 ventrals*. The olive-brown back is paler towards the tail, and there may be scattered black flecks on the sides of the forebody. Sometimes the body scales are black-edged. The white-yellow belly may have black streaks. An active, diurnal hunter that often moves with the forebody lifted. It eats various small vertebrates, including other snakes. Up to 30 eggs are laid in summer. A nervous snake, it swiftly retreats but will bite readily when caught. The venom may cause nausea, but is not dangerous.

Short-snouted Grass Snake *(Psammophis brevirostris)* 90-120 cm

This snake is easily confused with the Olive Grass Snake as it is similar in appearance and habits. However, it is a *smaller, more slender* species, and usually has *fewer than 164 ventrals*. The body is often striped, usually with a white 'stitch line' down the backbone and a lighter band on the flanks. The belly is white or yellow, often with a black spotted line on each side. Western individuals are more varied in colour, and spots on the side of the forebody are usually well-developed. It prefers dry grassland and savanna, where it hunts small vertebrates. From 4-15 eggs are laid in summer.

Western Sand Snake *(Psammophis trigrammus)* 90-110 cm

 The most slender sand snake in the region with a thin body. The *very long tail is more than half the body length* and has *132-155 subcaudals*. Body colour is cryptic olive to grey-brown, sometimes with a reddish-brown or yellowish-white lateral stripe. The off-white belly may have a grey to olive band down the middle. It is restricted to sparse bush in the western arid regions, but avoids open dune areas. Sand lizards and skinks are hunted during the heat of the day. Prey is not constricted, but simply swallowed alive. Like other sand snakes, a small clutch of elongate eggs is laid underground in summer.

Stripe-bellied Sand Snake (*Psammophis subtaeniatus*) 90-120 cm

 A beautiful, slender snake, but one that rarely stops to be enjoyed. The *bright yellow belly*, bordered by black and white stripes is unmistakable. The back is also striped, with a broad, black-edged dorsal band flanked by cream and brown stripes. It is often seen moving with its head up, alert for prey. It is a fast, active hunter of birds, lizards and mice in the open savannas of the northern regions. Due to its speed it is difficult to catch, although many are caught by birds of prey. From 4-10 elongate eggs are laid underground in summer.

Namib Sand Snake (*Psammophis leightoni*) 90-120 cm

 A brightly coloured, boldly striped snake with a dark, dorsal band that is often broken with a *conspicuous, thin, dashed line down the spine*. The *top of the head is spotted or barred*. The *anal scale is divided*. Like other sand snakes, it often forages with the head held up. Small lizards are chased, seized and swallowed alive. From 4-9 elongate eggs are laid in summer. The subspecies *P. l. namibensis* is shown. Other races, to the south and east, are less brilliantly marked.

Karoo Sand Snake *(Psammophis notostictus)* 80-100 cm

Very similar to the Namib Sand Snake, but usually *duller in coloration*. In the hand, it can be distinguished by having an *undivided anal scale*. It is one of the most common snakes of the Karoo and adjacent regions, and is often seen sunning on roads during the heat of the day. Lizards form the main diet and are caught after a quick chase. Quickly subdued by the mild venom, they are swallowed head first. At night it shelters in a rock crack or rodent burrow. A small clutch of 3-8 elongate eggs is laid in summer.

Dwarf Sand Snake *(Psammophis angolensis)* 30-40 cm

An elegant, tiny sand snake that is rarely seen, it has only *11 midbody scale rows* and a conspicuously *striped body*, with a broad brown-black dorsal stripe and a faint broken black stripe on the lower flank. The *dark brown head has 3 narrow crossbars* and *1 or several dark neck collars*. It forages among grass tussocks and fallen logs in lowveld savannas, where it feeds on small lizards and frogs. A small clutch of 3-5 elongate eggs is laid in moist soil.

Typical colour

Unmarked phase

 A small and robust snake. The tail is relatively shorter than that of other sand snakes (*61-81 subcaudals*). The silver-grey back usually has a *broad, black-edged, brown dorsal stripe*, and there is a similar stripe on the flanks. The head has *dark-edged, cream-yellow crossbars*. Some individuals lack stripes and are plain grey-olive. Restricted to cool, moist fynbos in the southern Cape coastal regions, with scattered populations in montane grassland further north. Lizards and frogs form the main diet, and 5-13 eggs are laid in midsummer.

Dwarf Beaked Snake *(Dipsina multimaculata)* 30-45 cm

This small, slender snake comes in a wide variety of ground colours that closely match those of the sands in the western rocky deserts. From 3-5 rows of dark, sometimes pale-centered, blotches occur on the back, and may fuse to form irregular crossbands. The neck has a dark V-shaped mark, and the distinct head has a *prominent hooked snout*. The *short tail has only 28-45 paired subcaudals*. It is a cryptic, slow-moving snake that ambushes small lizards from its hiding place in loose sand at the base of a bush or stone. From 2-4 elongate eggs are laid in summer. When threatened it may coil and hiss, pretending to be a small adder. Nonetheless, it is harmless.

Rufous Beaked Snake *(Rhamphiophis rostratus)* 120-140 cm

Like the Dwarf Beaked Snake, this species is characterized by a *prominent hooked snout*. However, it is a much larger, stout-bodied snake, and the *long tail has 87-118 paired subcaudals*. The body is a uniform warm yellowish to red-brown colour, sometimes with pale-centred scales. The head has a *distinctive dark brown eye stripe*. It shelters in burrows in the northern sandy bushveld, and eats a variety of small vertebrates. From 8-17 large eggs are laid in summer. It may hiss and strike when first caught, but tames well.

35

Bark Snake *(Hermirhagerrhis nototaenia)* 25-40 cm

A pretty, but secretive snake that shelters under loose bark or in hollow trees in the northern savannas. It is small and slender, with a *distinct, flattened head and large eyes with vertical pupils*. The *grey back has a dark dorsal stripe* flanked by, and sometimes fused with, a series of black spots. The dorsal pattern in the western population is more vivid. Small lizards, particularly Dwarf Day Geckos, form the main diet. It swallows its food as it hangs head down in the branches. A small clutch of 2-8 elongate eggs is laid in a tree hollow.

Bicoloured Quill-snouted Snake *(Xenocalamus bicolor)* 45-60 cm

Quill-snouted snakes are so bizarre they cannot be confused with any other snakes. They have *thin, very elongate bodies* and *quill-shaped heads with under-slung mouths*. The *minute eyes* have round pupils, and the tail is short and blunt. Coloration of this species is very varied, from striped, spotted, and mottled to uniform black. It is found in scattered populations in the sandy northern bushveld. It burrows in deep sand, searching for worm lizards that form its exclusive diet. From 3-4 eggs are laid in summer.

Cape Centipede Eater *(Aparallactus capensis)* 25-35 cm

Although very common in highveld grassland, this small, slender snake is rarely seen. It is characterized by a *small head with a rounded snout* and a *prominent black collar*. The body colour varies from red-brown to grey-buff, and the belly is cream-coloured. Most of its life is spent underground, in tunnels, rotting logs or rock piles, and particularly old termite nests. It searches for centipedes, which quickly succumb to the snake's venom, although this is completely harmless to humans. From 2-4 small elongate eggs are laid in summer.

Black Centipede Eater *(Aparallactus guentheri)* 30-45 cm

Although similar in build and habits to the Cape Centipede Eater, this species differs in having a blue-grey to black head and body, with *2 narrow yellow collars* on the neck. The chin and belly are off-white. In addition to centipedes, it also hunts scorpions. It is an East African species that just enters the subcontinent along the eastern escarpment of Zimbabwe. Breeding unknown.

Common Purple-glossed Snake *(Amblyodipsas polylepis)* 40-60 cm

 When the skin is freshly shed, the smooth scales of this black snake have an attractive purple gloss. Unlike the Natal Black Snake (page 39), with which it is easily confused, it has *15-31 paired subcaudals*. It is a stocky snake, with a blunt head and small eyes. There are *19-21 midbody scale rows*, and usually only 6 upper labials. It is found in the moister regions of the north-eastern savannas, and feeds mainly on burrowing reptiles, particularly blind snakes. Small prey is swallowed alive, whilst larger prey is constricted. Breeding unknown. It is a gentle snake and rarely bites.

Natal Purple-glossed Snake *(Amblyodipsas concolor)* 40-60 cm

 A thick-bodied, smooth-scaled snake of the eastern moist forests. The flattened head has a *blunt snout*, and the body is *uniform glossy black*, often with a purple sheen. It can only be differentiated from the Common Purple-glossed Snake in the hand. It has only *17 midbody scale rows*, *7 upper labials*, and *28-39 paired subcaudal scales*. It spends most of its time burrowing in leaf litter. Adults eat other snakes, although juveniles also eat lizards. Some females lay eggs while others give birth to up to 12 young in summer. Although it has large back fangs, its venom is not considered dangerous.

Bibron's Burrowing Asp *(Atractaspis bibronii)* 40-50 cm

 The *long, erectile front fangs* of this snake could cause confusion with vipers, with which the burrowing asps were long classified (as mole vipers). Many individuals are uniform black, above and below, whilst others have a prominent white belly. The short tail has *undivided subcaudals* and a *terminal spine*. It is usually found sheltering in rock piles, old termitaria, or under rotting logs, where it searches for other burrowing reptiles. Nestling mice may also be eaten. Up to 7 eggs are laid in summer. Although the venom of these bad-tempered snakes results in a painful bite, it has caused no known deaths.

Natal Black Snake *(Macrelaps microlepidotus)* 70-100 cm

 Like Bribron's Burrowing Asp, this burrowing black snake feeds on other burrowing reptiles. However, it also readily takes small mice and frogs. It *lacks erectile fangs*, but still has a toxic venom that has caused nausea but no deaths. It can be distinguished from burrowing asps by its larger size and fatter body, and has *no terminal spine* on the tail. It is endemic to the eastern coastal regions of South Africa, where it can be found burrowing in deep leaf litter in coastal bush, from Stutterheim to Zululand. The female lays up to 10 large eggs in the summer months.

Striped colour phase

 Unmistakable due to its bright colours, this small, slender snake is the region's most beautiful species. It is found in a number of different colour phases. Most snakes are patterned in black and yellow, and have a *bright red dorsal stripe* that may be broken into

Spotted colour phase

spots. Snakes in the Border region and Highveld are all black with a yellow dot on each scale, and a yellow dorsal stripe. It is found under stones, or rotting logs, and feeds mainly on other snakes and legless lizards. Up to 6 eggs are laid in December. The bite is painful but not fatal. It is now included with cobras in the Elapidae.

Striped Harlequin Snake *(Homoroselaps dorsalis)* 20-30 cm

 This minute, but elegant snake is easily recognized by its *very slender body*, which is *black with a lemon-yellow dorsal stripe*. The lips, flanks and belly are yellow-white. It lives underground in the grasslands of southern Mpumalanga and adjacent KwaZulu-Natal and Free State. Feeds mainly on thread snakes. Also included in the Elapidae.

Common Egg Eater *(Dasypeltis scabra)* 55-90 cm

 When disturbed, an egg eater puts on a great pretence of being dangerous. It gapes its mouth wide, revealing a *black lining*, and strikes readily. It may also make a hissing noise by rubbing *enlarged and keeled flank scales* against one another. It is in fact almost toothless, and completely harmless. The dirty grey body has numerous dark blotches and a *prominent V-shaped mark on the neck*, so resembling the Night Adder (see page 57). Found throughout southern Africa, it feeds exclusively on bird eggs. Up to 25 eggs are laid in summer.

Southern Brown Egg Eater *(Dasypeltis inornata)* 60-90 cm

 Easily distinguished from the Common Egg Eater by its *longer tail* and *uniform brown coloration*. It is restricted to open woodland in the eastern regions. Like other egg eaters, it has adapted to a strict diet of bird eggs. These are swallowed whole, and special throat 'teeth' (projections of the backbone) saw through the egg shell. The liquid content is then swallowed, and the collapsed egg shell is regurgitated. From 3-4 meals a year probably supply enough nutrition for growth and breeding. From 6-18 eggs are laid in summer.

41

Semiornate Snake *(Meizodon semiornatus)* 40-60 cm

A small, slender snake with *irregular black crossbars on the foreparts of the grey-olive body*. The *flat, black head* has relatively large eyes that are partially ringed in white and have round pupils. It inhabits the margins of pans and marshes, feeding on small frogs. During the dry season it shelters in hollow logs and under dead tree bark, where several individuals may be found together. It is known from isolated populations in Zimbabwe and Mozambique, extending southwards into northern Zululand. A few large eggs are laid in summer. *P

Marbled Tree Snake *(Dipsadoboa aulica)* 60-80 cm

A rare, secretive inhabitant of mature forest along the large rivers of the Lowveld and adjacent Mozambique and Zimbabwe. The *big head is finely marbled with white*, and the eyes have *vertical pupils*. Juveniles have a light brown back with 38-57 light, dark-edged crossbands. These decrease in size and number with growth. It shelters in bamboo and palm thickets, emerging at night to feed on geckos and reed frogs. Up to 8 small eggs are laid in summer.

Eastern Tiger Snake *(Telescopus semiannulatus)* 60-90 cm

 This distinctive, nocturnal species is thin-bodied, with an *obvious head and large eyes*. It has *19 midbody scale rows* and a *divided anal scale*. The *dull orange body has 22-50 dark blotches that are larger on the forebody*. Although mainly terrestrial, it regularly climbs into trees or house roofs to search for food. Small birds and lizards form the main diet, although mice and even bats are also caught. From 6-20 eggs are laid in moist leaf litter. When disturbed, this species bites readily and often, but the mild venom is not dangerous.

Beetz's Tiger Snake *(Telescopus beetzii)* 40-60 cm

 Similar in appearance to the Eastern Tiger Snake, but smaller and less aggressive. It has *sandy-buff coloration* and *more dark blotches* (30-39 on the body and 12-20 on the tail) *21 midbody scale rows* and an *undivided anal scale*. It is a rare species, sheltering in rock outcrops in the western deserts and scrublands. It is nocturnal, emerging at night to hunt lizards. The female lays a small clutch of 3-6 eggs in the summer months.

Spotted Bush Snake *(Philothamnus semivariegatus)* 80-100 cm

A beautiful, diurnal snake that hunts among bushes on rocky ridges or along river courses. The *slender body* has a *long tail*. A *lateral keel* runs on each side of the belly and tail. The *green body has black spots or crossbars on the foreparts* and may become grey-bronze towards the tail. It is an expert and speedy climber, and pursues lizards and tree frogs. A small clutch of 3-12 eggs is laid in midsummer. When confronted it inflates the throat to expose *vivid blue skin between the scales* and strikes readily. Despite this bluff it is harmless.

Western Green Snake *(Philothamnus angolensis)* 80-100 cm

More robust than the Spotted Bush Snake, this species usually hunts in reedbeds and marginal vegetation along major river courses. The body is *bright emerald green*, without black spots but with *scattered bluish-white scales*. It feeds on frogs and lizards, but will also take nestling weaver birds. Like the bush snake, it inflates the throat in threat, but has *black skin between the scales*. Up to 16 elongate eggs may be laid, and females sometimes nest together. *P

Natal Green Snake *(Philothamnus natalensis)* 70-100 cm

 Due to its *bright green body*, this snake is often mistaken for a juvenile Green Mamba. However, it *lacks fangs or venom* and has *keeled ventral and subcaudal scales* (which are reduced or even absent in southern populations). It has large eyes and *2 pairs of temporals on each side of the head*. An active and alert hunter of small frogs and fish. Lizards may also be eaten. Prey is not constricted, but simply swallowed alive. Up to 8 elongate eggs are laid in summer. When disturbed it inflates its throat in threat.

Green Water Snake *(Philothamnus hoplogaster)* 60-90 cm

Very similar in habits, appearance and distribution to the Natal Green Snake, and impossible to distinguish except in the hand. It has a *rounder head*, the *temporals are not paired*, and the *ventrals and subcaudals lack keels*. Some, particularly juveniles, may have black crossbands on the forebody. It is an active swimmer and hunts small frogs. Up to 8 elongate eggs are laid in early summer. Rarely attempts to bite and does not inflate the throat in threat.

45

Red-lipped Snake *(Crotaphopeltis hotamboeia)* 60-75 cm

Easily distinguished from all other local snakes by its *bright red-orange lips*. However, in northern populations this colour fades and the *glossy black temporal region* and *white flecks on the olive body* are better identification features. The large eyes have vertical pupils. It is an inhabitant of marshy areas and hunts small frogs. Up to 12 eggs are laid in early summer. Bad-tempered, it flattens the head to accentuate the red lips if threatened. Nonetheless, it is harmless, although the long back fangs can inflict deep punctures.

Barotse Water Snake *(Crotaphopeltis barotseensis)* 40-60 cm

This rare, poorly known snake is restricted to the Oka-vango Swamp and upper Zambezi River, where it hunts frogs at night in the papyrus swamps. It is very similar in appearance and size to the Red-lipped Snake, but has a *more elongate body and smooth, glossy scales in 17 rows at midbody*. The head and body are light grey-brown, with dark-edged scales; the belly is pale brown. Of gentle disposition, it rarely bites. Up to 8 eggs are laid in February.

Boomslang *(Dispholidus typus)* 120-160 cm

Female

One of the most characteristic snakes of southern Africa, absent from the drier western regions. Juveniles have bright emerald eyes, white throats and cryptic, twig-coloured bodies.

Male

Females remain drab olive, whilst males may become mottled in black and gold, or uniform bright green, rust-red or even powdery blue. All can be easily distinguished by the *very large eyes* and *oblique, strongly keeled body scales*. It is a dangerous but shy diurnal snake that hunts chameleons and small birds. Up to 25 eggs are laid in early summer. When disturbed it inflates the throat and will bite readily. The venom prevents blood clotting and death may follow in 1-3 days.

Twig Snake *(Thelotornis capensis)* 80-120 cm

The *very thin, elongate body, lance-like head,* and *cryptic coloration* make this snake unmistakable. The *large eyes* have *keyhole-shaped pupils*. The body scales are feebly keeled, and the *tail is very long*. The twig-coloured body is grey-brown with black and pink flecks and a series of diagonal pale blotches. The crown is uniform green or blue-green in the north east, heavily speckled in the south-east, and with a Y-shaped mark in the north. It is completely arboreal and hunts lizards and small birds which are swallowed as it hangs downwards among the branches. From 4-18 elongate eggs are laid in summer. Its venom may cause death from internal bleeding.

47

ELAPIDS (Family Elapidae)

Relatively large snakes with *large, hollow, front fangs that are not hinged*. The body is usually covered in smooth scales, and the head has large, symmetrical scales. Most lay eggs and many have powerful venoms that can cause death from paralysis.

Yellow-bellied Sea Snake *(Pelamis platurus)* 60-80 cm

The *bright yellow and black stripes, flat head,* and *oar-like tail* are unmistakable. It drifts in surface currents, ambushing small fish sheltering in floating debris. Sea snakes are common in the East Indies, but this is the only species that enters southern African coastal waters. Vagrants are washed south in the Agulhas Current from tropical seas around Madagascar. These snakes weaken in our cool waters and are eventually washed onto the eastern beaches by onshore winds. From 3-5 young are born in the surface waves at sea. The powerful venom causes paralysis, but no deaths have been reported.

Coral Snake *(Aspidelaps lubricus)* 40-70 cm

In southern populations, the *black-banded, bright orange body* is diagnostic. In northern Namibia, how-ever, adults are drabber and larger but can be identi-fied by a *large rostral scale on the nose* and only *19 smooth midbody scale rows*. A nocturnal hunter of semi-desert areas, it eats small lizards and rodents. From 3-11 eggs are laid in summer. When disturbed it rears the forebody, and spreads a very narrow hood. The venom is relatively mild, but has caused death. In captivity it feeds well, but remains bad-tempered.

Typical form

Lowveld form

 Less brightly coloured than the Coral Snake, this short, thickset snake has a much larger, *bulldozer-like rostral scale*. The scales are in *21-25 rows at midbody*, and become *heavily keeled on the hindbody*. The head and neck are usually black, with a *white throat band*. It burrows in loose sand and feeds on rodents and frogs. Up to 12 eggs are laid in summer. When disturbed it may rear and spread a narrow hood. If huffs and puffs a lot, but usually strikes with a closed mouth.

Sundevall's Garter Snake *(Elapsoidea sundevallii)* 50-80 cm

Adult

Garter snakes are notoriously difficult to identify. There is little difference in scale counts between species, and the brightly banded juvenile pattern may be present, absent or modified in adults. The *snout is slightly pointed. Juveniles have*

Juvenile

alternating bands of cream-pink and chocolate-brown, separated by narrow white rings. The latter persists in adults in the south, but elsewhere all bands fade resulting in slate-grey to black upperparts with a pink-buff belly. It is slow-moving and nocturnal, and eats rodents, lizards and even other snakes. Up to 10 small eggs are laid.

Boulenger's Garter Snake *(Elapsoidea boulengeri)* 50-60 cm

A slightly smaller species than Sundevall's Garter Snake with a *more rounded snout* and *3 lower labials* which touch the anterior chin shields. *Juveniles have a white head* and *12-17 narrow white bands* on the black back. In adults (over 20 cm) the pale bands darken in the centre to leave very narrow paired bands, which in large adults may disappear completely. It is a rare, shy snake that eats small vertebrates, including other snakes, and in captivity is prone to cannibalism. Up to 10 small eggs are laid in summer.

Günther's Garter Snake *(Elapsoidea guentheri)* 40-50 cm

This rare snake is restricted to miombo woodland in central and northern Zimbabwe. Like Boulenger's Garter Snake, the *snout is also rounded*, but *4 lower labials* touch the anterior chin shields. *Juveniles are black* with *16-20 white crossbands on the body* and *2-4 bands on the tail*. In adults these fade to leave narrow paired bands that may be absent in large specimens. Like other garter snakes, their diet is varied and includes other snakes. Up to 10 eggs are laid in late summer. Nothing is known of the venom.

Rinkhals *(Hemachatus haemachatus)* 90-120 cm

L. HOFMANN

This large, stout cobra is named after the characteristic *white throat band* (which may be divided into two). It differs from other cobras in having *keeled body scales* and being viviparous. *Juveniles are conspicuously banded*, with about 40 irregular, alternating black and tan bands which may fade with age. Adults on the Highveld are dirty grey-black. Adults in Zimbabwe and the eastern Cape may retain bright yellowish-orange bands. It is mainly nocturnal and hunts toads and small rodents in damp grassland. Up to 63 young are born in autumn. The toxic venom is potentially fatal. In defence it can spit venom up to 3 metres, or it may sham death.

Black Spitting Cobra *(Naja woodi)* 130-160 cm

This larger, diurnal race is found from southern Namibia south to Ceres. It is distinguished by its *black coloration* and its *221-228 ventrals*. Juveniles are grey with a black head and throat. It is an active hunter of rodents and lizards among the rock outcrops and dry water courses of the succulent Karoo. Up to 22 eggs are laid in summer. It spits venom willingly, but few bites are known. *R

Western Barred Spitting Cobra *(Naja nigricincta)* 120-140 cm

The *narrowly banded* pattern is superficially similar to that of the Rinkhals, but it has *7-21 smooth mid-body scale rows*. The head is broad with a rounded snout. The light grey to pink-brown body has 51-86 black bands on the back and 13-32 bands on the tail. There is a broad, dark throat band. It is a nocturnal hunter of frogs and rodents in the more well-watered regions of northern Namibia. Up to 20 eggs are laid in summer. It is capable of spitting toxic venom up to 2.5 metres, and bites result in extensive skin necrosis.

Mozambique Spitting Cobra *(Naja mossambica)* 90-130 cm

This small spitting cobra has a *blunt head* and *23-25 midbody scale rows*. The body is pink-grey to dark olive and each scale is *edged with black*. The belly is pinkish and may have irregular black crossbands or blotches on the throat. Its diet is varied, and mice, lizards, amphibians, and even grasshoppers are eaten. From 10-22 eggs are laid in summer. Due to its nocturnal habits, it is responsible for many snakebites in Zululand and the Lowveld, but fatalities are rare. When disturbed it spreads a broad hood and spits readily.

Forest Cobra *(Naja melanoleuca)* 180-210 cm

The largest and most impressive cobra of the subcontinent. The *slender body has 19 rows of glossy scales*. The head and forebody are yellow-brown, *heavily speckled with black*, whilst the tail is shiny blue-black. The *lower labials are often white with black edges*, particularly in northen populations. Its inhabits rainforest, and is restricted to northen Zululand and the eastern escarpment of Zimbabwe. It is fond of water and eats small vertebrates, including fish. Up to 26 large eggs are laid in summer. When cornered it will rear, spread a *narrow hood*, and bite. Bites are very rare, but serious. *P

 A slender, nervous cobra with a *broad head* and smooth, but *dull scales in 19-21 rows at midbody. Juveniles are dirty yellow, speckled with dark brown, and have a broad black band on the throat.* Adult coloration varies, from uniform yellow (Kalahari region) to dark mahogany (Namaqualand), although most specimens are heavily flecked in dark brown. The black throat band fades with age. It is an active, diurnal hunter. Small vertebrates, including other snakes, form the main diet and they commonly raid Sociable Weaver (*Philetairus socius*) nests in the Kalahari. From 8-20 large eggs are laid underground. It spreads a broad hood and bites readily, although it does not spit venom. The venom is highly toxic.

Anchieta's Cobra (Naja anchietae)

Snouted Cobra (Naja annulifera)

 Stout, large-headed cobras, characterized by having a *row of scales (suboculars) between the eye and upper labials*. The body is usually yellowish brown, with old specimens becoming blue black. The Snouted Cobra has a *banded phase*, with 7-9 broad yellow bands. A dark throat band is conspicuous in juveniles. They are nocturnal, emerging at dusk to hunt small vertebrates. Other snakes, particularly puffadders, are regularly eaten. From 8-33 large eggs are laid in summer. They spread a broad hood and bite readily, but do not spit. They may sham death when restrained. The venom causes death from respiratory failure.

55

Black Mamba *(Dendroaspis polylepis)* 200-300 cm

A *very large*, slender snake with a *coffin-shaped head* and smooth, but dull, scales. The body is dirty grey, sometimes olive, with black blotches on the pale grey-green belly. The *mouth lining is black*. An active, mostly terrestrial species, that feeds mainly on rats and dassies. These are pursued and repeatedly stabbed until succumbing to the very toxic venom. A small clutch of 12-18 large eggs is laid. Africa's most feared snake. When disturbed it rears the forebody, gapes widely, spreads a very narrow hood, and hisses. Heed the warning!

Green Mamba *(Dendroaspis angusticeps)* 180-220 cm

A smaller, more slender mamba than the Black Mamba. The body is *brilliant green*, and the *mouth lining is white*. Locally it is restricted to the forests of coastal Zululand and the eastern escarpment of Zimbabwe. It is arboreal, living in the upper forest canopy and searching for small mammals and birds. It is shy and rarely seen. Up to 10 eggs are laid in summer. Bites are very rare, and although not as dangerous of those of the Black Mamba, are still very serious.

ADDERS (Family Viperidae)

Characterized by large, hollow, *hinged front fangs* through which venom is injected. The body is usually stout with a short tail, and the head is wide, with *small, irregular scales on the crown* (absent in primitive night adders). Most adders are viviparous.

Rhombic Night Adder *(Causus rhombeatus)* 40-80 cm

A stout snake, although thin for an adder, with a *rounded snout* and *soft, feebly keeled scales*. There are *large, paired scales on top of the head*. The grey-pink body and tail have *20-30 dark, pale-edged rhombic blotches*, and there is a characteristic *dark V-shape on the back of the head*. It inhabits moist situations in the eastern regions, emerging at night to feed on toads. From 15 - 26 eggs are laid in summer. Although aggressive when first caught, it tames readily. The mild venom causes swelling but there have been no reported deaths.

Snouted Night Adder *(Causus defilippii)* 30-40 cm

Similar in build and coloration to the Rhombic Night Adder, but smaller and with *triangular dorsal blotches*. The *snout is pointed and upturned*. It is common in the Lowveld, and occurs in more arid savanna in Zimbabwe. It feeds almost exclusively on small amphibians. The female lays small clutches of 6-8 eggs during the summer months. The venom of this snake is mild and can cause only local swelling and pain.

 A large, thick-bodied, sluggish snake that has a short tail and a *triangular head covered in small scales*. The body is yellow-brown to light brown, with *numerous dark, pale-edged chevrons on the back*. Males are more brightly coloured than females. It is found throughout the region, except in deserts and high mountains. About 20-30 young are born in the late summer. A very dangerous snake that is responsible for many bites, although these are rarely fatal.

Gaboon Adder *(Bitis gabonica)* 80-120 cm

 Another large, heavy adder with a triangular head covered in small scales, and with a *pair of horn-like scales on the snout*. The *body has an attractive geometric pattern* of purple, brown and other pastel colours, whilst the *pale head has a thin, dark central line*. It is perfectly camouflaged when sheltering among leaf litter. Locally it is restricted to the coastal forests of northern Zululand and the escarpment forests of eastern Zimbabwe, where it is endangered due to habitat destruction. Up to 43 young are born in late summer. It is docile and rarely bites, but should be treated with respect. *V

Berg Adder *(Bitis atropos)* 30 - 45 cm

Eastern escarpment form

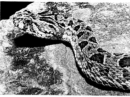

This small adder is found in isolated populations in the escarpment mountains, from Inyanga in Zimbabwe to the Cedarberg in the south-western Cape. Most are boldly patterned in grey and blue-black, but the colour varies between populations. The *head is elongate*, lacks horns, and has a *dark arrowhead on the crown*. The stout body is covered in *keeled scales in 29-33 rows at midbody*. Up to 15 young are born in autumn. It hisses and strikes readily but fortunately its venom is mild.

South-western Cape form

Plain Mountain Adder *(Bitis inornata)* 25-30 cm

A small, fat adder that has *27-31 midbody scale rows*. The *back is dull brown with fainter dark blotches*. The belly is dirty cream with blotches restricted to the sides. It is poorly known and apparently confined to the montane grassland of the Sneeuberg near Graaff-Reinet. It is active during the early morning and evening, and hides among small stones and grass tussocks to ambush passing lizards, particularly skinks and sand lizards. From 6-8 young are born in the late summer.

59

Many-horned Adder *(Bitis cornuta)* 30-50 cm

A medium-sized adder, easily recognized by the *blotched grey-black dorsal pattern*, and a *tuft of 2-4 horn-like scales above each eye*. Symmetrical dark markings on the crown may fuse to form an arrow-head shape. It shelters under rocks and in rodent burrows, ambushing rodents and lizards. In summer it is most active during the early morning and evening. Found mainly in the coastal regions of the western Cape and southern Namibia. From 4-10 young are born in late summer.

Horned Adder *(Bitis caudalis)* 25-35 cm

Northen Cape form

Easily recognized by the *single horn set above each eye*. The body is blotched and the background colour varies regionally, from light grey in Etosha to buff or orange-brown

Northern Province form

in the Kalahari, and grey-olive to light brown in the Karoo and Northern Province. The head has a *dark V-shape on the crown*. The creamy-white *belly is always unpatterned*. It is a common snake of the western arid regions. Lizards form the main diet. From 4-27 young are born in the late summer. The venom is mild and no deaths have been reported.

Péringuey's Adder (*Bitis peringueyi*) 20-25 cm

 Endemic to the Namib Desert, this species is easily identified by its small size and the *unusual position of its eyes, which are placed high on the flattened head*. The subcaudals are *smooth*, except for small keels towards the tip. The body is pale or reddish-brown with faint dark spots. It is known for its ability to sidewind, moving in smooth lateral curves that lift most of the body off the hot sand. It often shuffles into loose sand, leaving only the eyes exposed. From this position, small lizards are ambushed.

Namaqua Dwarf Adder (*Bitis schneideri*) 18-24 cm

 The smallest adder in the world, and similar in appearance and habits to Péringuey's Adder. However, its *eyes are placed on the side of the head* and *subcaudals are more strongly keeled*. It inhabits more compacted coastal sands in Namaqualand, where its habitat is endangered by strip mining for alluvial diamonds. Small lizards, and sometimes frogs, are eaten. From 3-4 small young are born in late summer. It will hiss and strike in defence, but its venom is very mild and there have been no serious injuries. *V

Desert Mountain Adder (*Bitis xeropaga*) 40-50 cm

 Living in the inhospitable mountains bordering the lower Orange River, this small, relatively slender adder is poorly known. It has a ridge above each eye, but these bear no horns. The *subcaudals are smooth* and the *belly is light grey, patterned with dark speckles*. The buff-grey *back has 16-34 dark-light bars*. It is a rock-dwelling species that does not shuffle into sand. Small rodents and lizards are eaten. From 4-5 young are born in the late summer. There have been no recorded bites, but this snake should be treated with care.

Lowland Swamp Viper (*Proatheris superciliaris*) 40-60 cm

 A medium-sized adder, with a robust body and *elongate head*. Although most of the head shields are fragmented, there is a *large scale above each eye*. The grey-brown body has 3 rows of blackish spots which are separated by yellowish bars. The *undersurface of the tail is yellow-orange*. Females grow larger than males. It is restricted to low-lying marshes in the flood-plain of the Zambezi River, where it lives in rodents' burrows, emerging at night to feed on small frogs. From 3-8 young are born in early summer.

LIZARDS (Suborder Sauria)

SKINKS (Family Scincidae)

A diverse family with numerous legless species. They have *shiny, overlapping scales* with an internal bony layer (osteoderms). This flexible armour protects them as they burrow. The *head has large, symmetrical scales* and the tail can be shed and regrown.

Giant Legless Skink *(Acontias plumbeus)* 30-45 cm

 This *very large* species is the largest legless skink in the world. Like other members of the genus, it *retains eyelids*. The *broad head* has an *elongate, steel-grey snout*. The *uniform black coloration* and large size are diagnostic. It prefers leaf litter in moist situations, where it burrows in search of invertebrates, although it also eats burrowing frogs and even nestling rodents. From 2-14 young are born in late summer. It grows slowly, but is easily kept in captivity, feeding on mince or pet food.

Short-headed Legless Skink *(Acontias breviceps)* 20-23 cm

 A medium-sized legless skink distinguished by its *broad head* with a *rounded snout*, and the *enlarged scales beneath the tail*. The body is olive to light brown, sometimes with a speckled appearance due to dark brown or black spots on each scale. The *belly is spotted*. It burrows underground in firm soils searching for earthworms and insect larvae. From 2-3 young are born in late summer.

Cape Legless Skink *(Acontias meleagris)* 20-26 cm

Widely distributed in the southern Cape, this *medium-sized* legless skink is often found in dry sandy soils beneath stones or dead trees. It has a *slender head and body*, with a *rounded snout* and *blunt tail*. The *lower eyelid is opaque*. The body is golden brown in colour, often with a dark spot on each scale in western populations. In the east these spots fuse to form longitudinal stripes. It is usually associated with richer soils. Up to 4 young are born in late summer.

Striped Legless Skink *(Acontias lineatus)* 15-17 cm

A *small*, *thin* legless skink with a characteristic *flattened, spade-like edge to the snout*. The *lower eyelid is transparent*. The *tail is flattened below*. Coloration is varied. Most have a yellow back with 4-10 fine longitudinal stripes. In the south-western Cape these may be replaced by fine spots. A few individuals from throughout the range are uniform black, or have black backs and pale bellies. It may be locally common in dry river valleys, where it burrows in loose sand at the base of vegetation. A single young is born in summer.

Striped Blind Legless Skink *(Typhlosaurus lineatus)* 14-18 cm

 Easily confused with the Striped Legless Skink, but almost blind. The *eyes remain only as black dots beneath the head shields.* A *long groove runs from the nostril* across the large, oval rostral. It is a slender species with *14 midbody scale rows.* Coloration is varied and uniform black specimens are common. The body is usually golden yellow with 2-8 fine stripes on the back. The belly may be plain or striped. From 1-2 large young are born in summer. *Re

Cuvier's Blind Legless Skink *(Typhlosaurus caecus)* 18-22 cm

 A *small* blind species with a slightly *flattened but rounded snout.* The front of the chin has a *heart-shaped mental scale* with a median cleft, and there are *207-230 ventrals.* The body lacks pigmentation, usually being *flesh-coloured,* often with a yellow-orange tinge on the back and an orange snout. It burrows in sand at the base of bushes in sparsely-vegetated coastal dunes in the western Cape, from Cape Town to Vredendal. Usually 1, possibly 2, babies are born in summer.

Gronovi's Dwarf Burrowing Skink *(Scelotes gronovii)* 10-13 cm

 A small, burrowing skink of the western Cape coastal region. It differs from Cuvier's Blind Legless Skink in having *well-developed eyes and eyelids*. It *lacks forelimbs*, but the *spike-like hindlimb has a single toe*. The *snout is flattened* and the *tail is slightly shorter than the body*. The silver-grey body has a vague brown band along the backbone, and the greyish-white belly is often heavily speckled. It is usually found beneath litter just above the highwater mark. From 1-2 young are born in March or April. *Re

Lowveld Dwarf Burrowing Skink *(Scelotes bidigittatus)* 12-15 cm

 A small species that *lacks forelimbs*, but has *2 toes on each of the small hindlimbs*. The *snout is rounded* and the *tail is as long or slightly longer than the body*. Each scale of the dark brown body has a dark spot, and there is a well-defined pale dorsolateral stripe. The *tail is metallic-blue*. It is found among dead leaves or under rotting logs. Known from scattered localities in the north-eastern Lowveld and northern Zululand. From 1-2 young are born in late summer.

Western Dwarf Burrowing Skink *(Scelotes capensis)* 9-10 cm

Within its range this small burrowing skink can be confused with nothing else. Its very *bright blue tail* and the functional, but *minute, 5-toed limbs* are diagnostic. There are *22 midbody scale rows* and the tail is as long as the body. Regenerated tails are usually much shorter than the body. The body is olive-brown with a coppery sheen, often with a lighter dorsolateral stripe on each side. It lives in moist soils beneath rocks. Restricted to the moister mountain slopes in Great Namaqualand and the Richtersveld. Breeding unknown.

Angola Burrowing Skink *(Sepsina angolensis)* 12-14 cm

A rare burrowing skink from Angola, whose range just extends into northern Namibia. It has a *stout body* with small but well-developed limbs, each of which has *3 toes*. The *snout is rounded*, the *lower eyelid is transparent*, and the *fattish tail* is about half as long as the body. The bronze, speckled colour of the body fades to silvery-pink on the tail. This species burrows in sandy soils in dry savanna. Breeding unknown.

67

Bouton's Skink *(Cryptoblepharus boutonii)* 8-10 cm

 This small, slender skink is locally restricted to a few coastal rock outcrops on the eastern seaboard, but occurs on small islands throughout the Indian Ocean. The large eyes have *immovable eyelids*, each with a transparent spectacle. The body is covered with *smooth, close-fitting scales in 26-29 rows at midbody*, and the tail tapers to a fine point. The *blackish-bronze back* has numerous pale spots on the flanks and legs. It is a diurnal species that forages on intertidal rocks. It swims readily, and dives into shallow pools to escape predators. From 1-2 eggs are laid in sand in summer. *V

Sundevall's Writhing Skink *(Lygosoma sundevallii)* 15-18 cm

 A small, fat-bodied skink that wriggles among leaf litter and loose sand. It has *movable eyelids* and a *fat tail*. The bronze body appears speckled due to a small spot on each scale. It is common in the northern savannas, where it feeds on termites and insects around rotting logs. From 2-6 soft-shelled eggs are laid under a stone or in an old termite nest in summer. The tail was once prized as a cure for snakebite.

Western Rock Skink *(Trachylepis sulcata)* 16-20 cm

Adult male, Karoo

Adult male, northern Cape

 This is one of the most common rock-dwelling lizards in the Karoo and western arid region. It is a *slender, flattened* skink, with a *window in each lower eyelid.* Females and juveniles are olive-brown with *6 dirty-gold stripes.* Breeding males vary from jet-black, with a heavily speckled throat in the Karoo to black with varying amounts of dirty bronze on the hindbody and tail in Namaqualand and Namibia. It shelters, sometimes communally, under large rock flakes. From 3-5 young are born in late summer.

Wedge-snouted Skink *(Trachylepis acutilabris)* 13-15 cm

 Easily confused with sand lizards, this small skink has *close-fitting, shiny scales* and a flattened snout with a *sharp edge to the upper lip*. The ear-openings are covered with *long, sharp lobes*. The light brown body has dark spots and white flecks that form short bands, and usually a pale dorsal stripe. It is a sit-and-wait predator, dashing from the shade of a small bush to seize insects. It is restricted to the western arid regions of Namibia. Breeding unknown.

Cape Skink *(Trachylepis capensis)* 20-25 cm

 A familiar garden lizard throughout South Africa. It is a fat, almost obese skink, with a *window in each lower eyelid* and *32-36 midbody scale rows*. The light brown body has *3 pale stripes, between which are numerous dark bars*. Some are olive-grey and almost patternless. It digs tunnels in loose sand and feeds on insects. Most females give birth to 5-18 young in late summer, although some lay eggs.

Western Three-striped Skink (*Trachylepis occidentalis*) 20-25 cm

 Very similar to the Cape Skink, but *more slender* and with relatively *longer feet*. There are *30-32 midbody scale rows*. The colour pattern is usually brighter, and patternless specimens are unknown. It occurs in the western arid regions, where it shelters in a short burrow dug in loose soil. Females lay eggs as well as give birth to live young.

Rainbow Skink (*Trachylepis margaritifer*) 18-24 cm

Male

 A common, brightly coloured skink with *42-44 midbody scale rows*. It forms large colonies inhabiting granite outcrops in the Lowveld and Zimbabwe. The females and juveniles

Female

have dark bodies with *3 bluish-white and electric-blue tails*. Breeding males become green-brown, with a pearly white spot on each scale and a *bright orange-brown tail*. From 6-10 eggs are laid in early summer.

Red-sided Skink *(Trachylepis homalocephala)* 15-18 cm

Male

A small, elegant lizard with a *boldly striped, shiny body.* In the breeding season, *males develop bright red flanks.* There is a small *window in each lower eyelid* and *28-30 midbody scale rows.* It forages in *Female* leaf litter around the base of coastal thickets from the Cape to the eastern escarpment. Isolated populations occur on inland mountains where rainfall is high. About 6 eggs are laid in early summer.

Hoesch's Skink *(Trachylepis hoeschi)* 17-19 cm

A medium-sized, slender skink with a longish, orange-brown tail. Like the Cape Skink, the body has *4 rows of dark blotches* that form irregular crossbars, with a pale lateral stripe on each flank. However, the markings are *lighter and the bars more numerous.* There is a *window in the lower eyelid and 32 rows of scales at midbody.* It inhabits rocky ground in the arid savannas of northern Namibia, where it feeds on wasps, beetles and other insects. Breeding unknown.

Variable Skink (*Trachylepis varia*) 12-16 cm

 A smallish skink with a rounded snout and *30-36 midbody scale rows*. Characteristically, there is a *bright white lateral stripe* and there are *3 keels on the scales beneath the toes*. The dark, reddish-brown back may have black spots and additional pale stripes. It is restricted mainly to the eastern regions and hunts in broken ground, climbing on boulders. Insects form the main diet and are captured after a short dash from cover. Whilst mainly live-bearing, some females in the Northern Province lay a single clutch of 6-12 eggs.

Variegated Skink (*Trachylepis variegata*) 11-14 cm

 Found throughout the western regions, this small, slender skink has *30-36 midbody scale rows* with only three keels on the scales. The brown body *lacks an obvious white lateral stripe*. It has only a *single keel on the scales beneath the toes*. It is active around the base of rocky outcrops, hunting small insects and spiders. Breeding males develop a rusty blush beneath the hind limbs and tail. Two to four babies are born in late summer.

Striped Skink *(Trachylepis striata* complex) 18-22 cm

Trachylepis wahlbergi

Widespread throughout the northern regions, these dull skinks tame readily and are common on garden walls and rockeries. *Midbody scale rows vary from 32-43* and colour differs from region to region. The dark-brown to black body may have bold white

Trachylepis striata

dorsolateral stripes (in the east) or numerous small, pale spots (in the south and central regions). The northern species has a pale grey back that may have faint stripes, a black band extends from the eye above the shoulder, and the breeding males develop yellow-orange throats. From 3-9 babies are born at a time, females in the south breeding only in summer, those further north breeding throughout the year.

Wahlberg's Snake-eyed Skink *(Panaspis wahlbergii)* 8-10 cm

A small skink, easily identified by its *immovable eyelids* and *grey-bronze body*, which may have 6 fine dark lines. The smooth body *scales are in 24-26 rows at midbody*. The belly is greyish-blue, except in breeding males when it turns pinkish-orange. It scuttles among grass roots and rotting logs, feeding on small insects. It survives for only 10-14 months. Females lay 2-6 eggs in early summer.

OLD WORLD LIZARDS (Family Lacertidae)

Small lizards with slender bodies, covered above with *small, granular scales*, and with *larger, quadrangular scales on the belly*. The head has large, symmetrical scales and the tail can be shed and regrown. At least 30 species are found in the region.

Shovel-snouted Lizard *(Meroles anchietae)* 10-12 cm

The unusual, *flattened snout with a sharp cutting edge* is diagnostic. The *flattened body has a silvery sheen* and the tail *may have a few black crossbands*. The body scales are small and granular, and the *toes of the long hindlimbs have a conspicuous fringe*. These allow it to race over hot sand. When disturbed, or to sleep, it dives into loose sand on a dune slipface. It is restricted to the windblown sands of the northern Namib Desert and eats insects and plant seeds. Several clutches of 1 or 2 large eggs are laid during summer.

Wedge-snouted Desert Lizard *(Meroles cuneirostris)* 13-15 cm

Very similar in appearance and habits to Anchieta's Desert Lizard, but with more *red-orange coloration* and an *unbarred tail*. Males have a blotched pattern. It is restricted to the southern Namib Desert, and only overlaps in range with Anchieta's Desert Lizard in the Lüderitz region. It is found on low, vegetated dune hummocks. There is no distinct breeding season, and 2-4 eggs are laid in soft sand.

Knox's Desert Lizard *(Meroles knoxii)* 15-20 cm

A small, active lizard that has a *rounded body*, but *lacks a sharp edge on the snout* and has only a *weak fringe along the toes*. The *supranasals are in contact*. Adult males have a yellow flush to the lips, throat and anal region. Juveniles are brightly striped in black and yellow. It lives in the coastal regions of the western Cape and Namaqualand, where it inhabits well-vegetated sand flats, sheltering in the shade of a grass tussock and making a quick dash to catch small insects. Up to 6 eggs are laid in summer.

Spotted Desert Lizard *(Meroles suborbitalis)* 16-18 cm

Namib form

Similar to Knox's Desert Lizard, but with the *supranasals separated*. The back colour is very variable and often matches the local ground colour. It is usually mottled pink-grey to slate, and some populations

Southern form

retain faded stripes from the juvenile colour pattern. Breeding occurs throughout the year and females lay 3-7 eggs at a time.

Reticulated Desert Lizard *(Meroles reticulatus)* 14-16 cm

 Restricted to the sparsely vegetated coastal dunes of the northern Namib Desert, this small lizard has a *weak edge along the upper lip*, but has a *conspicuous fringe to the toes*. It differs from the Wedge-snouted Desert Lizard in having a blue-grey body speckled with a network of grey spots. It burrows in mobile sand dunes and feeds on small insects. When disturbed it runs for a short distance before diving into loose sand. Up to 4 large eggs are laid in October.

Southern Rock Lizard *(Australolacerta australis)* 14-19 cm

 A graceful lizard with a *long tail*, a *well-developed collar*, and *small, granular and smooth scales on the back*. The dark olive back has rows of small pale spots that become bright orange on the flanks. It lives on the rugged summit slopes of the Cedarberg and the mountains to the south, where it basks and forages on large, vertical rock faces. From 5-7 eggs are laid beneath a sun-warmed rock slab in summer. *Re

Bushveld Lizard *(Heliobolus lugubris)* 16-20 cm

 A small lizard with a *well-developed collar*, and *small, keeled scales on the back*. The *long toes lack a fringe*. Adults have grey-tan to red-brown backs with vague crossbars and 3 pale dorsal stripes. The hatchling has a jet-black body with broken yellow-white stripes, and a sand-coloured tail. It walks slowly with a stiff-legged gait, and is believed to mimic oogpister beetles, which squirt a pungent acid when threatened. From 4-6 eggs are laid in loose sand in summer.

Common Rough-scaled Lizard *(Ichnotropis squamulosa)* 17-20 cm

 Recognized by its *small head* and small, *strongly keeled body scales*, this active lizard hunts for grasshoppers and termites in sandy clearings. It is cryptically coloured, the buff-brown body having narrow dark crossbands or blotches with long rows of pale spots. It is very short-lived, growing to maturity in 8-9 months and dying within 12-14 months after breeding. It digs branching burrows in soft sand at the base of bushes. From 8-12 eggs are laid in April to June.

Delalande's Sandveld Lizard *(Nucras lalandii)* 20-28 cm

 A *stout-bodied* lizard with a blunt snout and a *thick tail, that is almost twice the body length*. This serves as a fat store, but is readily shed. Unlike the Lined Sandveld Lizard, there are *no enlarged scales under the forearm*. The olive-grey *body has 8-10 irregular rows of white, black-edged spots*. In adults the black edges may fuse to form broken crossbars. The belly is white with numerous black spots. It spends long periods in a burrow beneath a rock slab or fallen log in grassland. When food is plentiful, it emerges to eat insects and flying termites. From 4-9 eggs are laid in summer.

Striped Sandveld Lizard *(Nucras taeniolata)* 25-28 cm

 A variable species, with a wide habitat tolerance. It is found in 3 isolated populations. The *tail is more than twice the body length* and there are *5-8 enlarged scales under the forearm*. The *body usually has 3 pale dorsal stripes*, which are separated from the spotted flanks by a broad black band. The tail is orange-brown in adults, but pinker in juveniles. It hides under rocks or fallen trees and searches for food at the base of bushes. Up to 7 eggs are laid in summer.

79

Western Sandveld Lizard (Nucras tessellata)

Karoo Sandveld Lizard (Nucras livida)

Truly some of the region's most beautiful lizards. The bright *black and white barred flanks* and *long red tail* of the typical species are unmistakable. However, the Karoo species lack the red tail and have yellow bars on the flanks. They are slow, terrestrial hunters of the western arid regions, emerging from burrows under stones. They are specialist feeders on scorpions and large beetles, which are dug from their retreats. They are unable to look for predators whilst slowly digging or searching for food. The bright red tail then helps to distract attention away from the vulnerable head and body. The female lays from 3-4 eggs in the summer months.

Spotted Sandveld Lizard *(Nucras intertexta)* 22-26 cm

 A large, graceful lizard with a long, orange-brown *tail that is more than twice the body length*. The light brown *back has numerous rows of white, black-edged spots*. The *belly is white, with dark flecks restricted to the sides*. The juvenile is more brightly coloured, with a coral red tail. It is a fast-moving, terrestrial species and is found throughout the arid Kalahari region, from northern Namibia to southern Mozambique. From 2-8 eggs are laid in midsummer.

Burchell's Sand Lizard *(Pedioplanis burchelli)* 14-16 cm

Juveniles

 This is one of the most common terrestrial lizards in the grasslands and fynbos of the Cape escarpment mountains. The *tail is only slightly longer than the body*, and a *distinct collar* and faint gular fold are present. *Juveniles* are *vividly striped* in black and gold, with a *blue flush to the tail*. *Adult*  Colours fade to cryptic browns in adults, although faint stripes may still be present. It prefers exposed bedrock and shelters in a chamber beneath a rock slab. A small clutch of 4-6 eggs is laid in soil under a rock slab in midsummer.

81

Namaqua Sand Lizard *(Pedioplanis namaquensis)* 14-17 cm

A small, slender lizard with the *tail noticeably longer than the body*. A *distinct collar* and faint gular fold are present, and there is an *enlarged crescent-shaped scale above the ear*. The *juvenile is vividly striped but has a pink-brown tail*. The adult becomes cryptically coloured in tans and browns, although in some regions adults retain faint stripes. The tail remains orange-brown. It is amazingly fast, scuttling between bushes in search of insects. It shelters in a burrow dug at the base of a bush. From 3-5 eggs are laid in November.

Plain Sand Lizard *(Pedioplanis inornata)* 14-16 cm

This rather drab, grey-brown lizard lives on the lower slopes of rocky outcrops in the western arid regions. The *tail is only slightly longer than the body*, and a *distinct collar* and faint gular fold are present. There is usually a series of *pale green spots on the flanks* and there is an *enlarged crescent- shaped scale above the ear*. It is active even on hot days, running in pursuit of small flying insects. It forms small colonies in suitable habitat. A small clutch of soft-shelled eggs is laid in moist soil beneath a rock slab in early summer.

Spotted Sand Lizard *(Pedioplanis lineoocellata)* 13-17 cm

 Very similar to, and easily confused with the Plain Sand Lizard. It differs in having a series of *pale blue spots on the flanks* and *lacks the enlarged crescent-shaped scale above the ear*. It is widely distributed, preferring rocky flats and broken ground. Adults may retain juvenile back stripes, particularly in Namaqualand. It is a sit-and-wait predator, grabbing small insects after a short dash from shaded cover. It shelters at night and in winter in a small chamber dug in soil beneath a flat stone. From 4-8 eggs are laid in midsummer.

Cape Mountain Lizard *(Tropidosaura gularis)* 14-18 cm

 Very similar to the Common Mountain Lizard, but grows larger and has a *pair of preanal scales*. In the breeding season males become vividly coloured, with bright green backs, bright orange flank spots and yellow speckles on the head. It prefers rocky slopes, foraging among mountain fynbos in gullies between rock outcrops. It shelters at night under a large rock slab on soil, and hibernates in this retreat during winter. It readily climbs rockfaces, and feeds on small insects attracted to the flowering heather. Breeding unknown.

83

Common Mountain Lizard *(Tropidosaura montana)* 12-15 cm

 This shy and secretive lizard may be locally common, but does not move far from thick cover and is therefore rarely seen. It has a short head, a relatively long tail, and spiny, overlapping body scales. A *collar is absent*, but a *faint gular fold* and a *single preanal scale* are present. The olive body has a dark streak along the backbone, a pale dorsolateral stripe, and a series of pale yellow spots on the flanks. These become bright orange in the breeding season. The tail has a blue-green flush. It lives and climbs among heathers and grass, where it is perfectly camouflaged. It is active in the early morning and evening, feeding on small insects. These are caught by a short dash from cover. A small clutch of 4-5 eggs is laid in summer, in a small chamber excavated beneath a stone or in a sun-warmed patch of soil. The eggs take about 34 days to hatch and hatchlings measure about 60 mm in length.

PLATED LIZARDS (Family Gerrhosauridae)

Endemic to Africa and Madagascar, more than half of the known species in this small family are found in southern Africa. They are diurnal, oviparous lizards, most having stout bodies, long tails and well-developed limbs. The body scales are rectangular and have osteoderms. *A prominent lateral fold is characteristic.*

Desert Plated Lizard *(Gerrhosaurus skoogi)* 20-25 cm

A large, unusual and endemic lizard of the dune seas of the northern Namib Desert. It has a *spade-like snout* that it uses to dive into loose sand. The *short tail* is only slightly longer than the body and tapers abruptly to a fine point. Adults are ivory-coloured with scattered maroon blotches. The chin, throat and lower chest are black. Juveniles are sand-coloured. It is found in small colonies, and feeds on wind-blown insects and dry plant debris. It may shelter for up to 24 hours under the sand. From 2-4 large eggs are laid in March to April.

Dwarf Plated Lizard *(Cordylosaurus subtessellatus)* 12-14 cm

This small, elegant lizard is easily recognized by the long, *bright blue tail* and *striped body*. It forages among small, succulent-covered rock outcrops, where it feeds on small insects. It frequently rests on its belly, raising its small feet off the hot ground. When threatened, it wriggles and is very difficult to catch without breaking the tail. The tail writhes after it is detached, thus distracting attention away from the lizard. From 2-3 eggs are laid in summer.

85

Yellow-throated Plated Lizard *(Gerrhosaurus flavigularis)* 25-35 cm

With its small head, *long tail* and *bright yellow, dark-edged, lateral stripe on each flank*, this elegant, medium-sized lizard is unmistakable. *Breeding males usually have a bright red throat*, although a few have blue throats. It is common in the eastern regions, often adapting to suburban gardens. It is terrestrial, living in a small burrow at the base of a bush. It is very alert and thus difficult to catch without breaking the fragile tail. From 4-6 eggs are laid in summer.

Kalahari Plated Lizard *(Gerrhosaurus multilineatus)* 30-35 cm

A large, handsome lizard, similar to the Yellow-throated Plated Lizard, but *more robust* and *lacking the yellow lateral line*. The brown body is speckled and adults develop a bright blue flush on the throat and flanks. It lives in a burrow dug at the base of a shrub in bushveld or Kalahari sandveld. With its powerful jaws it tackles large grasshoppers, beetles, and even scorpions. Breeding unknown.

Rough-scaled Plated Lizard *(Gerrhosaurus major)* 30-40 cm

C TILBURY

 A stout lizard, with a short head and large eyes. The *dorsal scales are large and very rough.* The *rounded, light brown body* may have dark speckling and a faint dorsolateral stripe. The chin and throat are pale cream. It favours large, soil-filled cracks in well-wooded rock outcrops, and feeds on large insects. It also eats soft fruit and flowers, and even small lizards. A few large, soft-shelled eggs are laid in summer. It tames easily and settles well into captivity.

Giant Plated Lizard *(Gerrhosaurus validus)* 40-60 cm

The largest plated lizard, exceeded in size only by monitor lizards. The *head and body are flattened* and the *dorsal scales are small and only faintly keeled.* Juveniles are black with a series of yellow spots on the back, and with bars on the flanks. These fade in adults, which have black backs with light yellow speckling. Breeding males have a pinkish-purple flush on the chin and throat. It is shy and difficult to approach, and lives on the upper slopes of large granite koppies in the northern regions. In addition to large insects, it also eats flowers, soft fruit and leaves. Up to 5 large eggs are laid in summer.

Short-legged Seps *(Tetradactylus seps)* 13-18 cm

 A small, *long-tailed* lizard with reduced, but fully formed limbs. The *head, body and tail are dark bronze, speckled with black,* and relieved only by cream spots on the upper lip and *irregular bars on the neck.* It is rarely seen, but locally common, favouring thick vegetation in moist clearings, either in coastal forest or on mountain slopes. Small insects such as bees and grasshoppers form the main diet. From 2-3 large eggs are laid in summer.

Common Long-tailed Seps *(Tetradactylus tetradactylus)* 18-24 cm

 This thin, *snake-like* lizard has only *4 toes on each of the very small limbs.* The *very long tail* is three times as long as the body. The back is olive with a pair of dark-brown dorsolateral stripes on each side. Short *black and white bars ornament the side of the neck.* It hunts in montane grassland or fynbos for small insects. At night it shelters in thick vegetation. A small clutch of 3-5 eggs is laid in summer.

GIRDLED LIZARDS (Family Cordylidae)

One of the few lizard families endemic to Africa. At least 40 species occur in southern Africa, including 3 unusual, snake-like species. The *rectangular body scales* are keeled and arranged in regular rows (girdles). The head is flattened and has symmetrical head shields with embedded osteoderms. The tail often has whorls of spines and is easily shed, but can be regrown. Most species give birth to a few live young.

Cape Grass Lizard *(Chamaesaura anguina)* 35-40 cm

An elongate, *snake-like* lizard with *minute limbs*, each with only *1 or 2 claws*. The body *scales are rough and strongly keeled*. The *tail is 3-4 times as long as the body* and easily lost, but rapidly regrown. It hunts insects in grassland, where the tan and brown body is perfectly camouflaged. A few small young are born in late summer.

Highveld Grass Lizard *(Chamaesaura aenea)* 30-35 cm

Very similar to the Cape Grass Lizard, but with reduced, yet perfectly formed feet, each with 5 *clawed toes*. The body scales are also slightly smaller, though strongly keeled. It is more boldly striped, and the flanks may have spots or a reddish-brown lateral stripe. It is restricted to the eastern escarpment grasslands, with an isolated population in the Winterberg. It gives birth to up to 12 babies during the summer.

A very common species of the coastal rocks and mountain summits of the Cape. The *body is usually mottled brown*, often with a pale dorsal stripe. The *body scales are strongly keeled* and the *tail is spiny*. It lives in dense colonies where there are suitable rock cracks in which to shelter. It is active in the early morning and evening, and on sunny days basks on prominent rocks, retreating to shaded cracks when temperatures get too hot. Grasshoppers and beetles form the main diet and these are caught by a short dash from cover. The adults are aggressive and fight to form social hierarchies. If threatened, it jams into a rock crack by inflating the body, and uses the spiny tail to protect its head. From 1-3 young are born in autumn. These may shelter near the mother's retreat during the first year. They mature in 1-2 years.

Highveld Girdled Lizard *(Cordylus vittifer)* 14-17 cm

Almost identical in appearance and habits to the Cape Girdled Lizard, this species differs in having a *row of elongate scales on the nape*, and *orange-red colouring*. A pale dorsal stripe may also be present. It is found among rocks on the Highveld and adjacent grasslands. It basks on rocks in the morning sun, and catches beetles and grasshoppers. From 1-3 young are born in autumn.

Peer's Girdled Lizard *(Cordylus peersi)* 15-17 cm

A *slender, uniform black* girdled lizard that climbs effortlessly over the smooth granite outcrops of Little Namaqualand. The *body scales are strongly keeled* and the *nostrils are swollen*. It is very visible when basking on the top of a large boulder, but is shy and retreats quickly to cover if disturbed. Small groups of 3-7 may share a large sun-warmed crack, particularly in winter. Insects and caterpillars form the main diet, and 2-3 young are born in autumn.

Karoo Girdled Lizard *(Cordylus polyzonus)* 20-25 cm

 This graceful girdled lizard can be distinguished from other girdled lizards by the *smaller, more numerous girdles*, and by a *black blotch on the side of the neck*. Dorsal coloration is variable. Juveniles are tan-coloured, chequered with dark brown. This may be retained in adults, although uniform black, orange, and light blue populations are known. It is common throughout the western arid region, and is found in loose colonies, each lizard inhabiting a rock crack in a boulder. It basks on a high point, making short forays to grab beetles or grasshoppers. From 2-3 young are born in autumn.

Warren's Girdled Lizard *(Cordylus warreni)* 20-30 cm

 This *large, flat* girdled lizard is found in small, scattered populations through the eastern escarpment mountains from Zimbabwe to Swaziland. It has a *dark back* and *varying amounts of yellow spotting and barring*. It favours deep cracks in large boulders that are sheltered by trees. It is shy and difficult to approach. It eats insects, snails and small lizards. From 2-6 young are born in late summer.

Tropical Girdled Lizard *(Cordylus tropidosternum)* 13-15 cm

 One of the few arboreal girdled lizards, this small, *round-bodied* species shelters in hollow logs and under bark. The *scales are very rough* and the back is lichen-coloured, with lighter flanks and belly. A dark lateral stripe extends from the neck along the side of the body, and the upper lip has a cream speckle. It feeds on moths, spiders, and winged termites. It lays down large fat reserves to tide it over winter. From 2-4 young are born in summer.

Armadillo Girdled Lizard *(Cordylus cataphractus)* 12-16 cm

Adult

 This *thick-set, flattened* girdled lizard is renowned for its defensive habit of rolling into a tight ball. It has a *broad head, very large, strongly keeled and often irregular girdles,* and *large spines on the tail.*

Defensive posture

The throat and belly are usually *heavily blotched.* It lives in large cracks in low rock outcrops in Namaqualand. As many as 5 or 6 individuals, apparently family units, may inhabit the same large crack. *V

Adult

Adult and juvenile

Due to its *large size* and *heavily spined body and tail*, this lizard is easily recognized. The *back of the head has a fringe of 4 large occipital spines*. It lives in long tunnels in rolling grassland. On sunny days it basks in the open, from which it gets the common name 'Sungazer'. It is long-lived, and 1 or 2 young are born every other year. Much of its habitat has been destroyed for maize farming and the surviving populations are threatened by illegal collecting. *V

Blue-spotted Girdled Lizard *(Cordylus coeruleopunctatus)* 13-16 cm

A small, slender girdled lizard, easily identified by its *dark body with scattered, enamel-blue spots*. The *dorsal scales are small* and not separated by granules. The flanks, snout, and sides of the head may be rust-coloured, whilst breeding males often have greenish-yellow to orange throats. It is restricted mainly to coastal rocks in the cool southern Cape, where it may form large colonies. It digs a tunnel in a soil- filled rock crack. From 2-3 young are born in autumn.

Graceful Crag Lizard *(Cordylus capensis)* 18-22 cm

Although very similar to the Blue-spotted Girdled Lizard, this species *grows larger and lacks blue spots*. It has long toes and a thin tail, and runs easily over the vertical walls of large rock faces. The body is basically grey-black, although pale yellow vermiculations may be present on the head and back. It lives on the rocky summits of the Cape fold mountains, and is shy and difficult to approach. A pair of adults often share the same retreat. It eats insects, particularly bees and wasps. From 2-3 young are born in autumn.

Cape Crag Lizard *(Cordylus microlepidotus)* 25-30 cm

Southern Cape form

 This species lives in solitary splendour in large rock cracks on the summits of the Cape mountains. Unlike girdled lizards, *the neck and flanks are covered in granular scales* and *the tail is less spiny*. Breeding males

Karoo form

have brightly coloured flanks (orange in the east, yellow inland). Juveniles and females are drabber, with grey-brown, blotched backs. It eats large beetles and grasshoppers and can inflict a painful bite. From 2-6, usually 4, young are born in late summer.

Drakensberg Crag Lizard *(Cordylus melanotus)* 20-28 cm

 The breeding males of this lizard are even more colourful than male Cape Crag Lizards. Although the two species are difficult to tell apart, their ranges do not overlap. It has large *powerful jaws*. Like other crag lizards, it has *small scales on the flanks* and a hinge on the upper surface of the skull. By clamping its jaws tight, the lizard can jam itself into a rock crack from which it cannot be removed. It is found in very dense colonies, and is the most common lizard on the inhospitable rocky mountains of Lesotho. Its diet of insects is supplemented with flowers and berries during summer. It hibernates in winter. From 3-5 young are born in late summer.

Cape Flat Lizard *(Platysaurus capensis)* 18-21 cm

The *very flattened bodies* and *beautiful colours of breeding males* make flat lizards unmistakable. The backs of all species are covered with granular scales and there are *scattered spines on the legs*. Females and juveniles in all species have black backs with 3 pale stripes. The male Cape Flat Lizard has a bright blue back and belly, with a coral-red tail when breeding. It lives in dense colonies on the granite rocks of the lower Orange River. Flying insects are eaten, along with flowers and small fruits. It shelters beneath a thin rock flake and lays 2 soft-shelled eggs in summer.

Sekukhune Flat Lizard *(Platysaurus orientalis)* 17-20 cm

Very similar in appearance to the Cape Flat Lizard but with a *transparent window in the lower eyelid*. Adult males have a *dark green head with 3 green stripes, a green back with scattered pale spots, and a brick red tail*. The throat and belly are blue. It is an exclusively rock-dwelling species of the Mpumalanga escarpment and adjacent ranges, with a localized population in Sekhukhuniland. It can be seen basking on rock faces at midday, but is shy and quick to retreat. It lays 2 soft-shelled eggs in summer.

 This variable species is distinguished by having an *opaque lower eyelid*. Numerous races are found on isolated rock outcrops from KwaZulu-Natal, through the eastern escarpment to southern Zimbabwe. It is usually observed basking or foraging on rocky boulders or bedrock. Females may lay several 2-egg clutches during summer.

Breeding males of the Natal Flat Lizard (above) have a *dark blue belly and a green gular region*. The race is restricted to rock outcrops in extreme northern KwaZulu-Natal and adjacent regions. Other races of this variable species are found on isolated rock outcrops throughout the Northern Province and adjacent eastern Zimbabwe.

The breeding male of the Lebombo Flat Lizard has a *black back and belly*, and the brick-red coloration of the tail extends onto the flanks. It is restricted to the Lebombo Mountains of eastern Swaziland and adjacent regions.

MONITORS (LEGUAANS) (Family Varanidae)

All monitors are similar in appearance, having long, flexible necks, well-developed limbs with strong claws, and a long tail that cannot be shed or regrown. The *head and body are covered in small, bead-like scales* that lack osteoderms. The tongue is long and snake-like. About 40 species are found throughout Africa and Australia, but only 2 of these occur in southern Africa.

Water Monitor *(Varanus niloticus)* 120-160 cm

Juvenile

Africa's largest lizard. The *head is elongate* and the *flattened tail is much longer than the body*. The black and yellow barred juveniles are much brighter than the adults. It forages for crabs and other aquat-

Adult

ic organisms. Juveniles eat mainly insects and small frogs. Up to 60 eggs are laid in an active termite nest.

Rock Monitor *(Varanus albigularis)* 90-130 cm

Similar to the Water Monitor, but smaller, with a relatively *shorter head and tail*. The *nasal region is swollen*. The drab, mottled colour is usually well sullied with old skin and ticks. It is a great wanderer in arid, rocky habitats. Large insects and millipedes are the main prey, although small tortoises and lizards are also taken. Up to 37 eggs are laid in a hole dug in the ground in spring. When cornered, it usually escapes into trees or rock cracks. It can inflict a painful bite, but is not poisonous.

AGAMAS (Family Agamidae)

A large, diverse family found throughout Africa and Australasia. The 10 southern African species are all very similar in appearance, being *plump, short-bodied lizards*, with *triangular heads* and *thin tails*. The head is covered in small, irregular scales and the big mouth has two fang-like teeth in the upper jaw. Males develop vivid breeding colours and defend a territory that may include several females. They are active, diurnal lizards, many of which feed predominantly on ants and termites.

Southern Rock Agama *(Agama atra)* 20-25 cm

Male (above), female (below)

Common on rocky outcrops throughout the southern regions. A *small dorsal crest* is present and extends along the backbone, becoming enlarged on the tail in individuals from Namaqualand. The breeding male develops a *bright blue head* and displays from a prominent rock. Females and juveniles are drabber, and more shy. Usually found in colonies. Lays up to 18 eggs, sometimes twice, in summer.

Anchieta's Agama *(Agama anchietae)* 15-20 cm

This small agama is restricted to the western arid regions and prefers small rock outcrops, where it is usually found singly or in pairs. It differs from the Southern Rock Agama in having *shorter toes, black-tipped scales on the soles of the feet,* and *scattered, enlarged spines on the back.* From 10-12 eggs are laid in summer.

Spiny Agama *(Agama hispida)* 15-20 cm

Although similar in general appearance and habits, the *small earhole* and *irregular pale blotches on the dark throat* distinguish this agama from the Ground Agama. The breeding male also has a *vivid, almost metallic, yellow-green head and body.* It displays from a prominent boulder. A pair may share a short tunnel dug into sandy soil at the base of a bush. It runs rapidly when disturbed and often uses a rodent burrow as a temporary shelter. From 10-15 eggs are laid in a hole in sandy soil.

Ground Agama (*Agama aculeata*) 15-22 cm

Western form

Northern form

 This species has a *relatively large earhole* and the *throat markings form a central network*. Enlarged spines on the back are arranged in regular rows. The body is cryptically blotched in browns, although *breeding males* develop *bluish heads*. It is widely distributed throughout the savannas and semi-arid regions of southern Africa and is most frequently seen feeding on a passing column of ants, or basking in a low bush. From 10-18 eggs are laid in a hole.

Namibian Rock Agama *(Agama planiceps)* 22-30 cm

 This large, graceful agama is easily recognized by its *large hind legs, small head, and long tail.* The juveniles and females have a grey-olive body with pale blotches and a *bright orange blotch behind the shoulder*. The dark head has symmetrical lemon-yellow blotches. The breeding male develops a metallic dull blue-purple sheen to the body, whilst the head, neck and throat are orange-red. The tail is olive-yellow at the base, turning to coral-red at the tip. It forms loose colonies on granite outcrops in northern Namibia. Insects, leaves and seeds are eaten. Breeding unknown.

Southern Tree Agama *(Acanthocercus atricollis)* 20-35 cm

Female

 This large agama differs from other agamas in having a *large black shoulder spot*. Females and juveniles have cryptic, lichen-coloured bodies, whilst breeding males develop bright ultramarine heads. When threatened it gapes widely to reveal a bright orange mouth lining. Commonly seen in the eastern bushveld clinging to a tree trunk. If feeds mainly on flying insects. From 4-14 eggs are laid in summer in a small hole in the ground.

Male

CHAMELEONS (Family Chamaeleonidae)

These unmistakable lizards are adapted for life in trees. They are mainly restricted to Africa and Madagascar. The compressed body and head are covered in *small, granular scales* and the large, *turreted eyes can move independently*. The *toes are bound in uneven bundles* enabling these lizards to clasp thin branches. The *tail is prehensile*, and cannot be shed or regrown. Insects are captured by the familiar *telescopic tongue*. At least 20 species, some unnamed, are found in the region.

Cape Dwarf Chameleon *(Bradypodion pumilum)* 13-16 cm

The *throat region is pale green to yellow*, with a *throat crest made of elongate scales that do not overlap*. The *tail is longer than the body*. The leaf-green body usually has an orange-red lateral strip. A beautiful and common resident of bushy fynbos and gardens in the south-western Cape. It is less solitary than many other chameleons, and 5 or 6 may be found in a single bush. Like all dwarf chameleons it gives birth to 6-10 young whilst clinging to a branch. The sticky membranes of the young adhere to leaves.

Drakensberg Dwarf Chameleon *(Bradypodion dracomontanum)* 12-14 cm

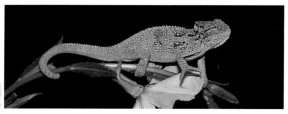

An inhabitant of evergreen kloof forests and upland scrub, this beautiful dwarf chameleon has a *recurved casque to the head* and a *throat crest of abutting scaly flaps*. As in all dwarf chameleons, the beautiful colours of the males only develop during mating and territorial displays. At other times they are coloured in cryptic browns and greys like females and juveniles. Breeding unknown.

Southern Dwarf Chameleon *(Bradypodion ventrale)* 12-14 cm

 This relatively large dwarf chameleon has a disjunct distribution, with isolated colonies in the western and eastern Cape. It has a *prominent casque on the back of the head*, and a *throat crest of large, overlapping, scaly flaps*. The *tail is shorter than the body* in both sexes. It lives in dense thickets in the east, and low coastal scrub in the west. Its colours make it difficult to see during the day, but at night it turns grey-white and is easily seen with a torch. It adapts well to urban gardens. Several litters of 10-20 young are born during summer.

Smith's Dwarf Chameleon *(Bradypodion taeniabronchum)* 8-11 cm

Male

 This small species is easily distinguished by its *black throat grooves*, which are prominent when the throat is inflated in threat. Adults are usually lichen-coloured, although males may be rust-red with maroon throat grooves. It is known from only a few small mountain ranges in the eastern Cape. It lives in protea bushes and low fynbos and feeds on insects. Litters of up to 13 young are born during summer. *E

Female

Setaro's Dwarf Chameleon *(Bradypodion setaroi)* 9-12 cm

A small species of low coastal dune forest in northern Zululand, but well adapted to urban gardens. The *casque is narrow and well developed*, but the *throat crest is reduced*. The *tail is longer than the body in males, but shorter in females*. The *gular region is light green with white throat grooves*. From 6-8 young are born in summer. *Re

Natal Midlands Dwarf Chameleon *(Bradypodion thamnobates)* 15-18 cm

This is one of the largest and most attractive dwarf chameleons, with a *large, recurved casque and a throat crest of long, overlapping scaly flaps*. A *pronounced dorsal crest extends along the back and onto the tail*. Males in display are dark blue-green with a cream or red-brown lateral patch. The cranial crest becomes horn-coloured. Like most chameleons, it is very aggressive and will threaten and chase off other chameleons entering its territory. Litters are very large, and up to 30 young may be born in summer. *Re

Flap-necked Chameleon *(Chamaeleo dilepis)* 20-24 cm

The only chameleon found in the savannas of the northern regions. It is characterized by *large occipital skin flaps* behind the head and a *crest of small, white scales on the throat and belly*. The body colour may vary from green to pale yellow or brown. It gapes widely in defence, revealing the orange mouth lining, and flattens the body whilst rocking from side to side. The female digs a long tunnel and lays up to 57 small eggs in summer.

Namaqua Chameleon *(Chamaeleo namaquensis)* 18-22 cm

The largest local chameleon, this terrestrial species inhabits some of the hottest and most desolate regions. It is ungainly, with a *large head, short tail*, and a series of *12-14 knob-like tubercles along the back*. It eats large numbers of grasshoppers and beetles, and will also eat small lizards and even snakes. It even forages along the hightide line. A nasal gland allows it to get rid of excess salt. Breeding occurs throughout the year, and 2 or 3 clutches comprising 6-22 eggs are laid in a burrow in sand.

GECKOS (Family Gekkonidae)

Widely distributed throughout the world, this is the most diverse lizard family in southern Africa, with over 60 species. Most local species have *immovable eyelids*, and wipe the eye clean with the tongue. Many have *elaborate toes*, with claws and expanded tips that have specialized scales covered with minute hairs (scansors). These allow them to climb seemingly smooth surfaces. Most are nocturnal, and have large eyes with complex pupils that open fully at night. They lay 1 or 2 hard-shelled eggs each year.

African Flat Gecko *(Afroedura africana)* 10-12 cm

As befits their name, flat geckos have *elongate, flattened bodies* with *3 pairs of scansors beneath the dilated, clawed toe-tips*. The *original tail is segmented* but carrot-shaped when regrown. The African Flat Gecko has a *pale-yellow to buff back* with 5-6 wavy, dark-brown bands that may break into blotches. It lives under thin rock flakes in scattered populations in Namaqualand and southern Namibia.

Transvaal Flat Gecko *(Afroedura transvaalica)* 10-12 cm

Superficially similar to the African Flat Gecko, this species has only *2 pairs of scansors beneath each toe-tip* and a browner body. It lives under flakes on rock outcrops. Despite the common and scientific names, it is found mainly in isolated populations in Zimbabwe. It is colonial, and up to 20 individuals may live in the same crack.

Hawequa Flat Gecko *(Afroedura hawequensis)* 12-15 cm

A very large and attractive flat gecko with a *stout body* and *3 pairs of scansors* under all toes, except the first. Males have *30-32 preanal pores in a curved row*. The original *tail is segmented*, and fat and leaf-like when regrown. It lives in narrow cracks among sandstone boulders in shady positions in the mountains of the south-western Cape. Two large eggs are laid in early summer. *Re

Karoo Flat Gecko *(Afroedura karroica)* 10-12 cm

Found singly or in pairs, this medium-sized species lives in fine rock cracks, among large sandstone out-crops or boulders in arid montane grassland in the eastern Cape. It has *3 pairs of scansors beneath the toes*, a *segmented tail*, and *no enlarged chin shields*. Preferred retreats are weathered sandstone flakes that catch the evening sun and are protected from seeping water. It emerges in the early evening to forage on rock faces for ants and small beetles. Two hard-shelled eggs are laid beneath a sun-warmed sandstone flake.

Tropical House Gecko *(Hemidactylus mabouia)* 12-15 cm

 This species is easily distinguished by its *large, flared toe-tips* that have *paired scansors* and a *large, retractable claw*. There are *10-18 irregular rows of weakly keeled tubercles on the back*. The body is pale grey with 4-5 wavy, dark crossbars that fade in light. Common in the Lowveld and KwaZulu-Natal, but rapidly expanding inland and along the eastern coastal region. Although normally found on trees where it shelters under bark or in hollow logs, it can adapt easily to houses, feeding on moths attracted to lights. Males are territorial. Communal nest sites may contain up to 50-60 eggs.

Tasman's Tropical House Gecko *(Hemidactylus tasmani)* 13-16 cm

 This large gecko is similar to the Tropical House Gecko, but *grows larger*, has *bigger, more strongly keeled tubercles* on the back, and *more conspicuous crossbars on the body*. It lives in rock cracks and caves in eastern and central Zimbabwe. A clutch of 2 eggs is laid.

Péringuey's Coastal Leaf-toed Gecko *(Cryptactites peringueyi)* 4-5 cm

This very small gecko has *leaf-shaped scansors, a sharp snout and numerous, enlarged tubercles on the back*. The red-brown body usually has a series of thin dark stripes, but these may fade. It is the only local lizard restricted to saltmarsh habitats. For a long time it was thought to be extinct but was rediscovered in 1992 in the Kromme estuary in the eastern Cape. It is a small, cryptic, nocturnal gecko, favouring rotting logs or matted vegetation. Clutches of 2 minute, hard-shelled eggs are laid throughout summer. *V

Striped Leaf-toed Gecko *(Goggia lineata)* 4-5 cm

This tiny, delicate gecko is the smallest lizard in the region. It often has a series of *conspicuous dark grey stripes on the pale grey back*, but these may be broken into irregular scallops. The body is round and the *toe-tips are flared*, with a single pair of large scansors. It is terrestrial and nocturnal, sheltering under rubble or among dead shrubs, particularly dried succulents. The diet is composed of small insects, particularly termites. Several clutches of 2 small, hard shelled eggs are laid in a moist, warm spot during summer.

111

Marbled Leaf-toed Gecko *(Afrogecko porphyreus)* 8-9 cm

This flat gecko is found throughout the southern Cape coastal regions. The feet are similar to the Striped Leaf-toed Gecko, and the *tail is round, unsegmented,* and longer than the body. The *back is smooth and marble grey in colour*, sometimes with a pale dorsal stripe. It is an adaptable species, living under tree bark, in rock cracks or even in suburban houses. It is not aggressive and as many as 24 individuals may share the same retreat. Suitable nesting sites, under tree bark or a stone on sandy soil, may be used by several females.

Small-scaled Leaf-toed Gecko *(Goggia microlepidota)* 9-13 cm

A large, attractive, but secretive gecko with *paired, leaf-shaped scansors* under each toe-tip. The *back is covered in minute, flattened scales*, from which it gets its name. The flat body is slate-grey with a blackish, reticulated pattern. It is restricted to the western Cape Fold Mountains from Cedarberg to Ceres. Here it lives in large rock cracks in the sandstone summit outcrops in mountain fynbos. It prefers large, shaded cracks, and is usually solitary. The breeding biology of this species is unknown.

Cape Dwarf Day Gecko *(Lygodactylus capensis)* 6-7 cm

 A delightful dwarf gecko that is often seen running on trees or garden walls. It has a *rudimentary inner toe*, while the *other toes have dilated tips, with large retractile claws and paired, oblique scansors*. The grey-brown body has a *dark streak from the snout to the shoulder*, which may continue as a pale lateral stripe. The throat is usually stippled with grey and the belly is cream coloured. It feeds almost exclusively on ants and termites. If disturbed it may freeze or quickly run behind cover which it keeps between itself and danger. Breeding is continuous and communal nest sites are common.

Methuen's Dwarf Day Gecko *(Lygodactylus methueni)* 7-9 cm

 A large dwarf gecko, whose toe structure is similar to that of the Cape Dwarf Day Gecko. The *large mental lacks lateral clefts*, and males have *9-11 preanal pores*. The brown to olive-grey body has *rows of pale-centred reddish-brown spots*. The *belly is yellow*, particularly at the rear and under the tail. It is endemic to the Woodbush Forest on the eastern escarpment, where it lives on rock outcrops. Its habitat is threatened by the development of pine plantations. Two eggs are laid under stones or loose bark. *V

Stevenson's Dwarf Day Gecko *(Lygodactylus stevensoni)* 6-7 cm

 This stout, dwarf gecko has a toe structure similar to that of the Cape Dwarf Day Gecko. It differs from this species in having a *more pointed snout*, a *blue-grey body* with *large, scattered black spots*, and *dark chevrons on the throat*. It is mainly rock-dwelling, although it may shelter under dead bark. It is restricted to well-wooded, rocky outcrops along the central Limpopo Valley. Breeding unknown.

Namaqua Day Gecko *(Phelsuma ocellata)* 6-7 cm

 All other day geckos are brightly coloured and found on Madagascar or other tropical Indian Ocean islands. It is therefore surprising that this small, drab, day gecko is endemic to the succulent, semi-arid wastes of Namaqualand. It has a *small inner toe*, which may cause confusion with dwarf day geckos. However, it has a *stouter build* and the *flared toe-tips have 7-8 undivided scansors and no claw*. The brown *body has numerous small dark and pale spots*. It prefers well-vegetated rocky hillsides. Small insects are eaten, and several clutches of 2 small, hard-shelled eggs are laid during summer. *Re

Festive Gecko *(Narudasia festiva)* 5-6 cm

 This very small, diurnal, flattened gecko is easily distinguished by the *slender, clawed toes that lack flared tips or adhesive scansors*. The brown body has numerous pale and dark blotches. The original tail is *bright orange*, but uniform grey when regrown. It forages on rocks in the western arid regions, but is restricted to southern Namibia. It is a very agile species, and catches ants and small flies on the rock faces and boulders of the mountains. Several clutches of 2 hard-shelled eggs are laid in a rock crack during summer.

Common Barking Gecko *(Ptenopus garrulus)* 6-8 cm

 Barking geckos differ from other local geckos in having *movable upper eyelids* and *short toes that lack scansors, but have a scaly fringe* to aid digging. In this small species the body colour varies from reddish-brown to greyish-yellow, being either finely speckled or having black crossbars. The male has an *orange or yellow throat*. Males call ('*ceek, ceek, ceek ...*') from the entrance of their burrows to attract mates. Their clicking chorus can be deafening during summer sunsets in the western arid regions. Several clutches of 2 (sometimes 1) hard-shelled eggs are laid during summer.

Carp's Barking Gecko *(Ptenopus carpi)* 7-10 cm

 A slender barking gecko, with *long legs and weakly fringed toes*. The *body scales are relatively large* and the *nostrils are not swollen*. The creamy-white back has fine orange-brown speckles and lines; there are also *3-5 dark brown crossbars on the back and 5-9 on the tail*. It burrows in more compacted soils on the flat, barren, gravel plains of the central and northern Namib Desert. The call is a monotonous series of slow, low-pitched clicks.

Koch's Barking Gecko *(Ptenopus kochi)* 8-10 cm

 A *large* barking gecko that has large, *bulging eyes*, *swollen nostrils*, and *minute body scales*. It has a rich reddish-brown body with dark speckles and an irregular row of light spots on the flanks. Adult males have sulphur-yellow throats and may have yellow flanks. It is endemic to the Namib Desert and builds a complex, branching burrow, up to a metre long, in the fine sands of interdune slacks or in the silt of dry river beds. Breeding unknown.

Giant Ground Gecko (*Chondrodactylus angulifer*) 13-16 cm

The largest local gecko, with a stout, cylindrical body, a large head, a short snout and large eyes. The *short toes lack scansors* and the back has *scattered keeled tubercles*. The *tail is segmented, with rings of enlarged tubercles, and is shorter than the body*. The back is usually pale orange to red-brown, sometimes with pale, dark-edged chevrons; the belly is pinkish-white. It inhabits gravel plains and sandy flats in the Namib Desert and Karoo. Nocturnal, it emerges from its burrow at night to feed on insects. When alarmed it walks stiff-legged with the tail arched, scorpion-like, over the back. Several clutches of 2 hard-shelled eggs are laid during summer.

Web-footed Gecko (*Palmatogecko rangei*) 10-12 cm

Easily recognized by its *large, jewel-like eyes*, this beautiful and bizarre gecko lives in the wind-blown sands of the Namib Desert. The large, flat head has *swollen nostrils*, and all the *toes are joined with webbing* and *lack adhesive scansors*. The tail is short and unsegmented. The semi-transparent body is fleshy-pink with dark reticulations, and there is a dark brown band across the snout. It spends the day in a tunnel dug in fine sand, and emerges to feed on crickets and spiders at dusk. It walks with a stiff-legged gait. Clutches of 2 large, hard-shelled eggs are laid from November to March. *P

117

Wahlberg's Velvet Gecko *(Homopholis wahlbergii)* 14-18 cm

A large arboreal gecko that shelters beneath bark or in holes in baobab trees, and may even inhabit old swallow nests or rock cracks on overgrown koppies and along river banks. The *dilated toepads have 8-12 unpaired, chevron-shaped scansors and small claws.* The *smooth back is covered in small, overlapping scales.* Males often have a pair of broad, black stripes on the grey back. The belly is usually a dirty-cream colour, sometimes with dark flecks. A pair of large, white eggs is laid in a rock crack or under bark.

Haacke's Thick-toed Gecko *(Pachydactylus haackei)* 12-15 cm

Like other members of the genus, this large, stout gecko is nocturnal and the *long toes have undivided scansors and rudimentary claws.* The rough, but delicate, skin tears easily if it is grabbed. Skin loss is a defence mechanism and, like a lost tail, is quickly replaced. The tail is segmented and there are *10-11 scansors under the middle toe.* The body may be light grey-olive or rich red-brown. It lives in rock cracks and emerges in the early evening to eat insects. Several clutches of 2 hard-shelled eggs are laid in summer.

Bibron's Thick-toed Gecko *(Pachydactylus bibronii)* 15-19 cm

The most familiar and widespread of the thick-toed geckos. The *back has numerous, scattered, enlarged, conical scales* that have a distinct longitudinal keel. It is a large, stout-bodied gecko, with a triangular head and powerful jaws. The grey to brown back has numerous scattered white spots, and 4-5 indistinct dark crossbands. It lives in rock cracks, under tree bark, and in houses, and forms large colonies. Several clutches of 2 hard-shelled eggs are laid during summer. Pugnacious, it is ever ready to give a painful bite.

Turner's Thick-toed Gecko *(Pachydactylus turneri)* 13-17 cm

This species is easily confused with Bibron's Gecko, but has *smooth tubercles on the back*. It is restricted to the arid western regions, and is usually only found on large rock outcrops. It shelters in rock cracks, but is less gregarious than Bibron's Gecko, usually occurring singly or in pairs. From 2-3 clutches of eggs are laid in sand in a rock crack in spring and summer. They hatch in 60-80 days.

Van Son's Thick-toed Gecko *(Pachydactylus vansoni)* 9-12 cm

 This species has a *cylindrical body* and the toe tips are only slightly expanded, with *4-6 scansors*. The back is brown with *narrow, white, black-edged stripes*. A terrestrial species, found from sea-level in Mozambique to the top of the Soutpansberg in the Northern Province. Several clutches of eggs are laid during summer.

Rough-scaled Thick-toed Gecko *(Pachydactylus rugosus)* 9-10 cm

 This small, round-bodied gecko is easily recognized by its *very rough, almost spiny skin*. The olive-green-brown *back has 4, creamy, wavy crossbands*. There is often a white stripe on the side of the head. It is restricted to the western arid regions where it lives under tree bark or among debris along dry river courses. When threatened it behaves like a scorpion, standing stiff-legged and arching the banded tail high over its back. It also gapes to reveal the pink mouth lining. Several clutches of eggs are laid during summer.

120

Velvety Thick-toed Gecko *(Pachydactylus bicolor)* 9-11 cm

 An elegant, small gecko with a *flattened head and body*, and *smooth, granular scales on the back*. Hatchlings are unusual, with *jet-black bodies bordered with white bands on the hips and neck, and an orange tail*. Like the hatchling sandveld lizards (pages 79-81), it may imitate the noxious oogpister beetle. The adult becomes buff-coloured with irregular brown blotches and bars. The eye has a golden ring and the upper lip is white. It lives in fine cracks in small, shattered rock outcrops, and feeds on small moths and spiders. Breeding unknown.

Weber's Thick-toed Gecko *(Pachydactylus weberi)* 9-10 cm

Although similar to the previous species, this gecko has *numerous rows of small tubercles* on the back and the hind legs. These may be golden in colour, giving an almost jewelled appearance. The juvenile is golden-brown with 3-4 cream, dark-edged crossbands, whilst the *tail has bright alternating black and white bands*. The body crossbands fade in the adult, but the banded tail is retained. It is very common in suitable habitat, but rarely seen. Usually only a single individual, occasionally a pair, inhabits a crack. Breeding unknown.

Austen's Thick-toed Gecko *(Pachydactylus austeni)* 7-9 cm

This species inhabits the sparsely vegetated coastal dunes of the western Cape. The *body is cylindrical*, the snout short, and the large eyes have a *conspicuous white or bright yellow eyelid*. The *body scales are smooth and granular*. Coloration is varied, ranging from pale grey to dark brown with scattered dark and pale spots that may fuse to form irregular bars. There are only 3-4 scansors beneath the toe-tips. It lives in a small burrow dug in sand, under stones and debris, and emerges at night to forage for small insects.

Speckled Thick-toed Gecko *(Pachydactylus punctatus)* 6-8 cm

This beautiful, docile gecko is terrestrial and lives in the poorly vegetated sand flats of the Kalahari and Namib regions. Although *more slender*, it is very difficult to separate from Austen's Gecko as it also *lacks enlarged tubercles on the back* and has only *3-4 scansors under the toe-tips*. Fortunately their ranges only overlap in the Richtersveld, where this gecko is always light orange in colour, with small, scattered pale spots. Elsewhere its coloration varies from pale-grey to purple-brown. It lives in burrows, and is usually found under stones on sand.

Spotted Thick-toed Gecko *(Pachydactylus maculatus)* 6-9 cm

 The most common terrestrial gecko in the eastern Cape, becoming rarer elsewhere. It has a *fat body* with a rounded snout, and *only small tubercles on the back*. The body is always pale grey with *four rows of black spots* that may be separated by white interspaces or fuse to form irregular stripes. A *black eye stripe* passes from the snout to the back of the head. It lives under debris, in rotting logs, or in old termite nests. Small insects and spiders form the main diet. From 3-4 clutches of 2 hard-shelled eggs are laid in summer.

Banded Thick-toed Gecko *(Pachydactylus fasciatus)* 9-11 cm

The hatchling differs greatly from the adult. It is yellow-cream with 2 wide *chocolate-brown bands on the body* and up to 10 bands on the tail. In adults, the dark bands fade in the centre to form *dark-edged, pale brown bands*. The cylindrical body has up to 18 regular rows of *enlarged and keeled tubercles*, and there are *4-5 scansors beneath the toe-tips*. It lives under calcrete slabs on sandy soils in northern Namibia. Hatchlings appear in April and May.

Ocellated Thick-toed Gecko *(Pachydactylus geitje)* 6-8 cm

This is another small gecko with a confusing array of colour patterns. The grey-brown to dark brown body usually has small, scattered, dark-edged pale spots, although these may fade in mountain populations. The *cylindrical body has smooth scales*, and there are *4-5 scansors under the toe-tips*. In the moister, western regions it is terrestrial, hiding among debris and under stones. Inland and further east it is restricted to mountains and is found more often in rock cracks. Two small eggs are laid every 6-8 weeks during summer.

Marico Thick-toed Gecko *(Pachydactylus mariquensis)* 8-10 cm

A small, *slender, thin-legged* gecko with a short snout, and smooth body scales. As with most terrestrial species, it has only *3-4 scansors under the toe-tips*. The *cylindrical, unsegmented tail* is usually slightly shorter than the body. The grey to pinkish-buff back has 5-6 wavy, reddish-brown, dark-edged crossbands. A pale dorsal stripe may be present. It is found throughout the southern arid region, with scattered populations in Namibia. By day it shelters in a burrow in sandy soil, and emerges in the evening to hunt small insects. Several clutches of 2 hard-shelled eggs are laid during summer.

Namib Day Gecko *(Rhoptropus afer)* 8-10 cm

 This is the most common of a group of 5 diurnal geckos of the arid Namib region that have *long legs and toes*, with a *reduced inner toe*. The *toe-tips are flared with undivided scansors.* This species has *5-6 scansors under the middle toe.* The back has small, rounded scales, and the grey, dappled body is perfectly camouflaged. However, the *lower surfaces of the legs, tail and throat are bright yellow.* Individuals 'flag' each other by lifting the tail to reveal the bright yellow colour. It favours broken ground, where it darts rapidly between boulders. Ants and small beetles form the main diet. Several clutches of 2 hard-shelled eggs are laid during summer.

Barnard's Day Gecko *(Rhoptropus barnardi)* 6-8 cm

 A small day gecko that lives in semi-desert in northern Namibia. It differs from the Namib Day Gecko in having *8 scansors under the middle toe* and *slightly keeled scales* on the back. The *dappled body has a red flush* and it lacks the yellow underparts. It lives in regions of higher rainfall, preferring small rock outcrops and ridges. Clutches of 2 eggs are laid in a rock crack in winter, and communal nest sites may contain over 200 old and new eggs.

AMPHISBAENIANS (Suborder Amphisbaenia)

These are very unusual reptiles that for a long time were thought to be lizards. About 12 species are found in the northern parts of the region. They are the most specialized burrowing reptiles, capable of driving tunnels through compact soils. They spend all their lives underground, feeding on insect larvae and other invertebrates. All local species *lack limbs, eyes, and external ear openings*. The scales of the cylindrical body are arranged in rings, giving them an appearance similar to worms. The head usually has a specialized cutting edge for digging. They may lay eggs or give birth to live young.

Kalahari Round-headed Worm Lizard (*Zygaspis quadrifrons*) 16-20 cm

This small worm lizard has a *rounded head* that lacks a cutting edge, and between *198-242 body annuli*. The body is uniform purple-brown above, with a lighter belly. It inhabits sandy scrub and bushveld in the northern regions, where it feeds on insect larvae and termites.

Cape Spade-snouted Worm Lizard (*Monopeltis capensis*) 25-30 cm

A large, *uniform pink* worm lizard with a *large spade-shaped snout, with a horizontal cutting edge* used to dig tunnels. Just behind the throat are 4-6 enlarged, elongate scales. It digs in the deep, red sands of the southern Kalahari, extending along the Limpopo River valley to southern Mozambique. It is only seen during excavations, or when uncovered by floods. From 1-3 young are born in late summer.

CROCODILES (Order Crocodylia)

These remnants of the dinosaurs' rule are now endangered throughout the world. All are aquatic and the 23 species are distributed throughout the tropical regions, with 3 African species. Only 1 species occurs in southern Africa.

Nile Crocodile (*Crocodylus niloticus*) 250-350 cm

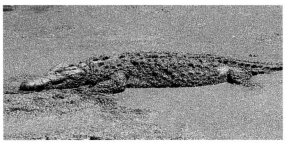

One of the largest living crocodiles that may exceed 1 000 kg in weight and nearly 6 metres in length. Like birds, the eye has an *extra eyelid* that sweeps away dirt. The eyes and *valved nostrils* are placed high on the head. The *hind feet are webbed* and the *long tail*, which cannot be shed, *has 2 raised dorsal keels*. Adults are a dull olive with a yellow or cream belly. Hatchlings are brighter, with irregular black markings and a straw-yellow belly. The female digs a nest hole on a sandbank, within which she lays 16-80 oval, hard-shelled eggs. Incubated by the sun, they hatch in about 85 days. The sex of the hatchlings depends on the incubation temperature; females are produced at low temperatures (26-30°C) and males at higher temperatures (31-34°C). The parents protect the nest during development, and also assist the hatchlings to and from the nest. Growth is slow, and maturity occurs in 12-15 years at 2-3 metres. Hatchlings and subadults live in marshes and backwaters, and feed mainly on insects and frogs. Adults eat fish, mammals and birds. Viable populations are now mainly restricted to game reserves. *V

CHELONIANS (Order Chelonia)

A very ancient group characterized by the presence of a *protective shell*. This consists of horny outer scutes (that may be lost in some aquatic forms) which cover a strong, bony layer.

SIDE-NECKED TERRAPINS (Family Pelomedusidae)

These primitive chelonians retract the head sideways. They are restricted to the southern continents.

Serrated Hinged Terrapin *(Pelusios sinuatus)* 30-45 cm

A large terrapin that has a *hard, domed shell with a distinct hinge* at the front of the plastron. This closes to protect the head. The rear of the shell is *serrated* and there are *keels along the backbone*, particularly in juveniles. The shell is black, except for a *yellow, angular-edged blotch in the centre of the belly*. It is common in large rivers and pans in the eastern region, where it is often seen basking on floating logs. From 7-13 soft-shelled eggs are laid in spring.

Pan Hinged Terrapin *(Pelusios subniger)* 13-18 cm

A small, hinged terrapin with a *rounded, smooth, dark brown shell* and yellow centres to the belly scutes. It shelters underground during droughts, and many are damaged by winter fires. It feeds on small frogs and invertebrates caught in shallow pans. In defence it withdraws into the shell and empties a foul-smelling liquid from its bowels. Small clutches of 3-5 eggs are laid throughout summer.

Marsh Terrapin (*Pelomedusa subrufa*) 20-30 cm

 The most common and widespread terrapin in southern Africa. The *hard shell is very flat* and has *no hinge* on the plastron. The *neck is withdrawn sideways into the shell*. There are 2 soft tentacles on the chin. It inhabits pans, vleis, and slow-moving rivers throughout southern Africa. When the pans dry, it burrows into moist soil. It eats almost anything, including small birds coming to drink. From 10-30 soft-shelled eggs are laid on a sandbank following summer rains.

SOFT-SHELLED TERRAPINS (Family Trionychidae)

These are unusual terrapins with only 3 toes on each foot and a soft, leathery shell. A bony layer is still present beneath the skin. They are restricted mainly to Asia and North America, with only 5 species occurring in Africa.

Zambezi Soft-shelled Terrapins (*Cycloderma frenatum*) 35-50 cm

 A large species with a *very long neck and a 'snorkel-like' nose*. The *hindlimbs are protected by flexible flaps* when withdrawn into the shell. Locally it is restricted to the Save River system of the Mozambique flood plain. It grows to 14 kg, and uses its strong forelimbs to dig in soft mud for snails and mussels. These are crushed by the strong jaws. Clutches of 15-22 hard-shelled eggs are laid in summer.

LAND TORTOISES (Family Testudinidae)

These advanced chelonians *withdraw the head backwards into the shell*. The hind feet are elephant-like and they walk on the tips of the heavily armoured forefeet. All are terrestrial, laying hard-shelled eggs. Fourteen species occur in southern Africa.

Greater Padloper *(Homopus femoralis)* 10-14 cm

A small tortoise, but the largest of the padlopers. The *shell is not hinged and has a small nuchal scale* and *paired gulars,* and usually only *11 marginals.* The forelimbs have only *4 toes* and a *large tubercle is present on each buttock.* The flattened shell is olive to rich red-brown, often with a wide black margin on each scute in juveniles. Males have longer tails than females but *lack a concave belly.* It inhabits the highveld and rocky montane grassland. From 1-3 oval, hard-shelled eggs are laid in summer.

Parrot-beaked Tortoise *(Homopus areolatus)* 7-10 cm

A smaller relative of the Greater Padloper, that also has only *4 toes on the forefeet.* However, the dorsal scutes of the shell often have *depressed centres* and shell abnormalities are common. The *beak is strongly hooked* (hence the common name) and the *nostrils open high on the snout. Buttock tubercles are absent.* Breeding males have bright orange noses. Rarely seen, it remains within cover to avoid predation by crows, jackals, and other predators. From 2-3 small eggs are laid in a small hole dug in sandy soil.

Boulenger's Padloper *(Homopus boulengeri)* 10-14 cm

 Another smaller species, but with *5 toes on the forefeet*. The *flattened shell has a rounded bridge* and usually *12 marginals*. The *beak is only weakly hooked* and there are *no buttock tubercles*. The shell varies in colour from olive to rich red-brown. Males are smaller than females and have a pronounced hollow belly. It is very secretive and favours rocky ridges in the Karoo, where it shelters under large rock slabs. It is usually seen on cool summer days when thunderstorms approach. From 1-2 eggs are laid in a small hole.

Speckled Padloper *(Homopus signatus)* 6-9 cm

Western Cape form

 The world's smallest tortoise, with a flattened, light-brown shell, heavily speckled and with serrated edges. The

North-western Cape form

shell in the southern race has smoother edges and a rich orange-red colour. There are 5 toes on the forefeet, usually *12 marginals*, and *buttock tubercles are present*. Males are smaller than females and have a pronounced hollow belly. It lives among the granite outcrops of Little Namaqualand, and emerges in the early morning to feed. Usually 1 egg is laid in a small hole. *Re

Angulate Tortoise *(Chersina angulata)* 15-25 cm

The only local tortoise that has an *undivided gular scute beneath the throat*. This is larger in males and used in combat to overturn opponents. A *nuchal scute is present*, the *carapace has no hinge*, and there are *no buttock tubercles*. The shell is light straw-yellow in colour with dark edges to the scutes. Old adults become smooth and dirty straw in colour. Some, particularly from the western Cape, have bright red bellies (the 'rooipens' form). Males are larger, and have a pronounced 'peanut' shape and a hollow belly. They are found throughout the Cape coastal regions, extending inland into succulent and broken karroid veld. The female lays a single, large egg.

Leopard Tortoise *(Geochelone pardalis)* 30-45 cm

The largest tortoise in southern Africa, easily distinguished by the *lack of a nuchal scute at the front of the shell*. The *gulars are divided*, the *carapace has no hinge*, and there are *2-3 buttock tubercles on each side*. Hatchlings are bright yellow, each scute having 1-2 black spots. Adults become darker and heavily blotched or streaked. Very old tortoises are almost uniform dark grey. Males have longer tails and hollow bellies. It is found throughout most of the region. Up to 6 clutches of 6-15 eggs are laid in summer.

132

Geometric Tortoise *(Psammobates geometricus)* 8-12 cm

An endangered species, restricted to coastal renoster-bosveld in the south-western Cape. Only 2 000-3 000 specimens survive in a number of special reserves. It is the only tent tortoise found in fynbos. The *shell is high and domed*, with *only slightly upturned rear margins*. The marginal scutes along the bridge are higher than they are broad. There is a small nuchal and a *single axillary*, but *no buttock tubercles*. The shell scutes usually have bright, radiating yellow and black rays. Females are larger and have smaller tails than males. A clutch of 2-4 eggs is laid in spring and hatches in late summer. *E

Kalahari Tent Tortoise *(Psammobates oculiferus)* 8-12 cm

The *low, domed shell, with strongly serrated front and rear margins*, is characteristic. The *nuchal is broad and often divided*. Like other tent tortoises, each shell scute has a radiating pattern of 6-10 dark rays on a tan or pale brown background. *Buttock tubercles are present*. Males have longer tails, more conical scutes on the back, and hollow bellies. The shell was often used by bushmen to make snuff boxes. It is found throughout the Kalahari region, where it feeds on small succulents and grasses. A clutch of 1-2 eggs is laid in summer.

133

Tent Tortoise *(Psammobates tentorius)* 8-12 cm

An attractive tortoise that comes in a wide variety of shapes and colours. The shell usually has radiating black and yellow rays on each scute, although some specimens are uniform brown. The shell shape may be flattened or domed, with raised or flat scutes. The *shell margins are not serrated*, and unlike the Geometric Tortoise, the *scutes along the bridge are broader than high*. It occurs throughout the Karoo and southern Namibia. Males are much smaller than females. A few eggs are laid in summer.

Bell's Hinged Tortoise *(Kinixys belliana)* 17-20 cm

A medium-sized tortoise which, when adult, has a characteristic *hinge in the rear of the carapace*. This closes to protect the hind feet and tail. The *carapace is slightly domed* and the *scutes have a radiating pattern of dark bands*. Adult males have a hollow belly. Locally it is restricted to moist savanna and thickets of the eastern coastal plain. It emerges in the early morning and evening to feed. From 2-7, exceptionally up to 10, eggs are laid in summer.

Natal Hinged Tortoise (*Kinixys natalensis*) 8-14 cm

A small species in which the *hinge is poorly developed* and restricted to the marginals. The *supracaudal is frequently divided*, the *beak has 3 cusps*, and the *plastron is not concave in males*. Scutes on the back and belly have *broad concentric light and dark zones*, although these may fade in old individuals. It inhabits dry, rocky areas, and hibernates from May to September. Breeding unknown. *R

Spek's Hinged Tortoise (*Kinixys spekii*) 13-18 cm

Another small species, with a *flattened shell* that allows it to seek refuge in rock cracks and hollow logs. The *beak has only 1 cusp*, the *carapace has a well developed hinge*, and the *tail ends in a spine*. The breeding male has a hollow belly. It spends the dry season in woodland, moving to more open savannas to feed on small flowers after the summer rains. It also readily eats fungi and snails, and especially millipedes. A small clutch of 2-4 eggs is laid in summer.

SEA TURTLES (Superfamily Chelonioidae)

Sea turtles have front feet which are modified into *flippers*, and they *cannot withdraw the head or feet into the shell*. They feed in the tropical seas, but return to sandy beaches to lay their soft-shelled eggs. Five species are found in the region's coastal waters, of which 2 nest on protected beaches in northern Zululand.

Green Turtle *(Chelonia mydas)* 98-120 cm

A non-breeding visitor to the eastern and western coasts, where it enters shallow estuaries to feed on sea grasses and jellyfish. The hard shell is smooth, with *non-overlapping scutes*. There are *12 marginals* on each side, which are smooth in adults. The *front flippers have a single claw*. Females are usually darker in colour than males. It is slow-growing, taking 10-15 years to mature. It is threatened by pollution, collection of eggs and slaughter for its meat. *V

Hawksbill Turtle *(Eretomochelys imbricata)* 60-90 cm

A non-breeding visitor to the east coast. The hard shell has *thick, overlapping scutes*. There are *12 marginals* on each side, those at the rear being *very serrated*. The *front flippers have 2 claws*. It is a relatively small turtle that feeds on corals and urchins. These are prised from the bottom with the *hooked beak*. Many millions have been killed for their shells, to provide the famous 'tortoiseshell' used in fashion. They breed in Madagascar and Mauritius. *V

Loggerhead Turtle (*Caretta caretta*) 70-100 cm

A frequent visitor to the east coast that breeds on protected beaches in northern Zululand. It is a large turtle, with a big head and an elongate shell that tapers at the rear. The scutes are *smooth* (keeled in hatchlings), *non-overlapping*, and each limb has *2 claws*. Both adults and young are brown. It hunts around reefs and rocky estuaries. The strong jaws are used to crush crabs, molluscs and sea urchins. Females come ashore on dark nights and lay up to 500 eggs at 15 day intervals. Up to 500 females nest each year in Zululand. *V

Leatherback Turtle (*Dermochelys coriacea*) 130-170 cm

R BOYCOTT

Found around the whole coast, with breeding sites on the northern Zululand beaches. It is the *largest* sea turtle, easily recognized by the *pliable, rubbery shell*, which has *12 prominent ridges*. The young are blue-grey and have long flippers. It is a specialist feeder on jellyfish, travelling the ocean currents in search of its prey. It may dive to over 300 metres, spending up to 37 minutes underwater. Females come ashore at high tide on moonless nights between November and January. They dig a shallow pit and lay clutches of 100-120 eggs, up to 6-9 times in one season. Juveniles hatch after 70 days and head out to sea. *V

137

Where to look

Reptiles are found everywhere. Even the cold summits of the Lesotho mountains are home to at least four lizard and three snake species. Lizards, particularly geckos and lacertids, however, reach their greatest diversity and abundance in the western arid regions. The sparse ground cover here allows them to bask and catch insects easily, and the many rock outcrops provide retreats for numerous rock-dwelling species.

Many snakes and other reptiles are tropical, widely distributed in east and central Africa. These reach their southern limit in KwaZulu-Natal, and include snakes such as the Gaboon Adder, Green Mamba, and Forest Cobra. Temperate species, on the other hand, including the Berg Adder, Rinkhals and Many-spotted Snake, are adapted to cool, moist climates. They are common in the south-western Cape, but further north are restricted to isolated populations on cool mountain summits.

Lizards are easier to find and to watch than snakes: they are more common, most are diurnal and they are very visible, as they regularly search for food. Skinks, agamas and flat lizards are common on rock outcrops in the northern savannas and western deserts.

A number of lizards have adapted well to urban parks and gardens. The Tropical House Gecko is very common in houses along the east coast, as is the Striped Skink in Highveld towns. Many gardens in the south-western and eastern Cape have thriving dwarf chameleon populations which are most easily observed after dark.

Tortoises are very common in the southern Cape. Because of the size differences in the species here there is no competition for food as different species feed on different plants. Terrapins are mostly restricted to the wetter northern savannas, and only the Cape Terrapin is found throughout the region.

Habitat

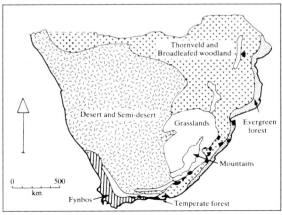

MAJOR VEGETATION ZONES OF SOUTHERN AFRICA

The map of the region (above) shows the major vegetation zones mentioned in the text. Reptile distribution is contained by and, in many cases, restricted to specialized habitats within these zones.

Glossary of terms

Arboreal Living in or among trees.
Bridge The side of a chelonian shell where the carapace joins the plastron.
Carapace The upper surface of a chelonian shell.
Caudal Pertaining to the tail.
Chelonian A shield reptile (tortoises, terrapins and turtles).
Clutch All the eggs laid by a female at one time.
Cryptic Hidden or camouflaged.
Diurnal Active during the day.
Dorsal Pertaining to the upper surface of the body.
Dorsolateral The upper surface of that part of the body bordering the backbone.
Ectotherm An animal, including all reptiles, that obtains its body heat externally by basking in the sun.
Femoral Pertaining to the upper part (thigh) of the hindlimb.
Granular Small, usually non-overlapping, scales.
Gular Pertaining to the throat region; a plate on the plastron of a chelonian shell.
Hinge A flexible joint in the shell of some chelonians.
Keel A prominent ridge, occurring on the back of some chelonians, and on the scales of some lizards and snakes.
Mental A scale on the head of a reptile.
Nocturnal Active at night.
Nuchal A scute at the front of the carapace of a chelonian shell.
Occipital Pertaining to the back of the skull.
Osteoderm A body scale containing a bony layer.
Plastron The lower surface of a chelonian shell.
Rostral Pertaining to the rostrum (nose); a scale at the front of a nose of a reptile.
Scansor Specialized scales found on the toe-tips of many geckos. These are covered in thousands of minute hairs that catch in cracks and allow the gecko to climb vertical surfaces.
Scute An enlarged horny plate on a chelonian shell.
Squamate A scaled reptile, including snakes, lizards, and amphisbaenians.
Temporal Pertaining to the side of the head behind the eye.
Tubercle An enlarged scale on the body of a lizard.
Ventral Pertaining to the lower surface of the body.
Vestigial Being smaller and of a simpler structure (a remnant) than in an evolutionary ancestor.

Colour patterns

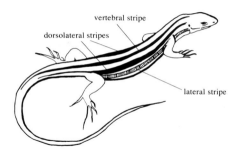

vertebral stripe

dorsolateral stripes

lateral stripe

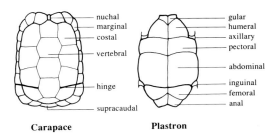

Carapace

Plastron

Head scales of a colubrid snake

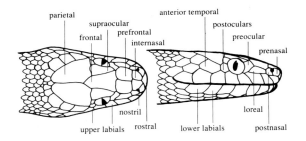

Further reading

Auerbach, R 1988. *Reptiles and Amphibians of Botswana*. Mokwepa Consultants, Gaborone.

Boycott, RC and Bourguin, O 1988. *The South African Tortoise Book: A guide to Southern African tortoises, terrapins and turtles*. Southern Book Publishers, Johannesburg.

Branch, WR 1998. *Field Guide to the Snakes and other Reptiles of Southern Africa*. rev ed Struik Publishers, Cape Town.

Branch, WR (ed) 1988. *South African Red Data Book: Reptiles and Amphibians*. South African National Scientific Programmes Report No. 151.

Branch, WR 1991. *Everyone's Guide to the Snakes of Southern Africa*. CNA, Johannesburg.

Broadley, DG 1990. *FitzSimons' Snakes of Southern Africa*. Jonathan Ball and Ad Donker Publishers, Johannesburg.

Jacobsen, N 1985. *Ons Reptiele*. CUM-Boeke, Roodepoort.

Marais, J 1992. *A Complete Guide to the Snakes of Southern Africa*. Southern Book Publishers, Johannesburg.

Patterson, R and Bannister, A 1987. *South African Reptile Life*. C Struik Publishers, Cape Town.

Pienaar, U de V, Haacke, WD and Jacobsen, N 1983. *The Reptiles of the Kruger National Park*. The Trustees of the National Parks Board, Pretoria.

Pooley, A 1982. *Discoveries of a Crocodile Man*. William Collins Sons and Co Ltd, Johannesburg.

Index

141